The Ethics of Climate Engineering

T0225375

This book analyzes major ethical issues surrounding the use of climate engineering, particularly solar radiation management (SRM) techniques, which have the potential to reduce some risks of anthropogenic climate change but also carry their own risks of harm and injustice. The book argues that we should approach the ethics of climate engineering via "non-ideal theory," which investigates what justice requires given the fact that many parties have failed to comply with their duty to mitigate greenhouse gas emissions. Specifically, it argues that climate justice should be approached comparatively, evaluating the relative justice or injustice of feasible policies under conditions that are likely to hold within relevant timeframes. Likely near-future conditions include "pessimistic scenarios," in which no available option avoids serious ethical problems. The book contends that certain uses of SRM can be ethically defensible in some pessimistic scenarios. This is the first book devoted to the many ethical issues surrounding climate engineering.

Toby Svoboda is an assistant professor of philosophy at Fairfield University. He has published in journals such as *Environmental Ethics, Environmental Values*, and *The Journal of Moral Philosophy*. He is the author of *Duties Regarding Nature: A Kantian Environmental Ethic* (Routledge, 2015).

Routledge Research in Applied Ethics

The Ethics of Climate Engineering

Solar Radiation Management
and Non-Ideal Justice

Toby Svoboda

Routledge
Taylor & Francis Group

LONDON AND NEW YORK

First published 2017 by Routledge

2 Park Square, Milton Park, Abingdon, Oxfordshire OX14 4RN

52 Vanderbilt Avenue, New York, NY 10017

Routledge is an imprint of the Taylor & Francis Group, an informa business

First issued in paperback 2020

Library of Congress Cataloging-in-Publication Data
A catalog record for this book has been requested

ISBN: 978-1-138-20483-6 (hbk)
ISBN: 978-0-367-59505-0 (pbk)

Typeset in Sabon
by Apex CoVantage, LLC

To Kate and Arizona

Contents

Acknowledgements

Several individuals provided helpful feedback on draft chapters of this book. For that I thank Daniel Callies, Frank Jankunis, David Morrow, Christian Uhle, and two anonymous reviewers. I acknowledge permission from the White Horse Press and the Center for Environmental Philosophy to reuse some material from two articles previously published in *Environmental Values* and *Environmental Ethics*, respectively. Finally, I acknowledge a Summer Research Stipend from Fairfield University, which supported work on this project during the summer of 2016.

Introduction

This is a book about justice, anthropogenic climate change, and the prospect of climate engineering (CE). More specifically, it is a book about the justice (or injustice) of using solar radiation management (SRM) in response to climate change. SRM is a category of CE, which we may define as the large-scale, intentional, technological modification of the global environment in order to address climate change.[1] A CE technique counts as SRM if it would seek to engineer the climate by increasing planetary albedo, such as by injecting sulfate aerosols into the stratosphere, thereby reducing the quantity of solar radiation absorbed by the planet. If they worked, SRM techniques could offset (to some degree) the planetary warming driven by the increasing concentration of atmospheric greenhouse gases due to anthropogenic emissions. A second category of CE, carbon dioxide removal (CDR), includes techniques that would reduce atmospheric carbon dioxide, such as via direct air capture and subsequent sequestration in geological formations or in the deep ocean. My focus in this book is SRM, although I will have occasion to refer to CDR techniques at several points.

I begin with the assumption that climate change is fundamentally a moral issue,[2] largely in virtue of the injustice it could bring for some present and future parties. I also assume that we ought to take much more ambitious action regarding climate change than we currently are taking—for instance, in the form of more aggressive mitigation of greenhouse gas emissions. The foregoing are not controversial views within the climate ethics literature. Unfortunately, it is clear that current commitments (e.g., those of the Paris Agreement) to address climate change are unlikely to be sufficient to avoid climatic change that is likely to carry injustice for some. It is also clear that some relevant parties have been unwilling to commit themselves to courses of action that are likely to be sufficient for this end. So far, it appears that (full) climate justice has been politically infeasible, and there is little reason to believe that this will change in the near future. Instead, we find ourselves accepting climate policies that, although presumably better than nothing, fall far short of what we ought to do. As Posner puts it, "You can have justice or you can have a climate treaty. Not both."[3]

The question is how to proceed when justice is politically infeasible. Two options immediately suggest themselves. First, we might work to change political conditions such that just policies become politically feasible, such as through various forms of activism. In effect, this is to adapt our social-political world to the demands of justice. Second, we might instead adapt justice to the demands of our social-political world, working to see what aspects of justice can survive (and to what extent) within the constraints of what is at present politically feasible. These two approaches roughly map onto what is called "ideal" and "non-ideal" justice, respectively. Now the second option might seem obviously worse than the first. After all, justice is inherently a normative concept, and so we might think that it should be immune to considerations of political feasibility. Indeed, most ethicists working on climate change have taking an ideal approach to climate justice, although recently more attention has been paid to non-ideal climate justice.[4]

But there are cases in which a non-ideal approach is called for, such as situations in which there is not sufficient time to achieve substantive change through activism, or situations in which the prospects of successful activism are very dim. In scenarios like that, we undeniably face the question of what we ought to do *given* the (perhaps deeply unjust) realities of our social-political world. Plausibly, this is the type of situation we find ourselves in with respect to climate change. I assume that many of us have obligations of ideal justice that require us (among other things) to accept climate policies designed to make substantial cuts in emissions, even if those policies carry significant costs to ourselves (e.g., in the form of a carbon tax). Yet it is difficult to deny that many of are not complying with our obligations to do this. Whether we like it or not, this puts us in the realm of non-ideal justice: given this fact of non-compliance on the part of some, well-intentioned parties must ask themselves what should be done. One danger of aiming for ideal justice under such circumstances is that we risk achieving little or nothing, whereas we might have achieved some imperfect degree of justice had our targets been less than ideal. So perhaps the idea behind Posner's dictum needs to be revised. It is not the case that we must choose to pursue *either* justice or a politically feasible climate policy. Instead, we might pursue some degree of justice within the constraints of political feasibility. In short, in some cases it may be that we can have both (non-ideal) justice and a politically feasible climate policy.

An Overview

This book is composed of six chapters. In the first, I address some ethical issues regarding the outcomes of various climate policies that we might adopt, including some involving SRM. We might be tempted to adopt a simple form of utilitarianism in assessing the ethics of climate policies, preferring that policy which delivers the best ratio of benefits to harms, aggregated across persons. Such an approach faces two important sets of problems. First, there is the issue of uncertainty. At best, our knowledge of the outcomes of some potential policy would be probabilistic in nature. There are, of course, methods for dealing with such uncertainty, but a greater problem is posed by the fact that we may be uncertain of the probabilities of certain outcomes, with different studies providing divergent probability density functions. That is, we may have cases of "deep" uncertainty. Second, a simple utilitarian approach risks overlooking the importance of how harms and benefits are distributed across persons. Intuitively, it matters a great deal (and in an ethical sense) whether climatic harms are borne predominantly by the rich or the poor, low emitters or high emitters, present or future generations, and so on. The aim of this chapter is not to argue that consequentialist (or, more specifically, utilitarian) approaches to climate ethics must fail, but rather to motivate the importance of justice. Although justice-based approaches to climate ethics also face the problem of uncertainty, I will argue in chapter three that procedural justice can help us navigate this problem. Moreover, most justice-based approaches are inherently concerned with the question of how harms and benefits are shared across persons, and they provide tools for assessing policy options in terms of their likely distributions.

The next two chapters respectively address distributive and procedural justice, taking a non-ideal approach to each. In chapter two I lay out my approach to distributive justice and explain its relevance to SRM. We immediately face the problem that, as a matter of political and moral theory, there are many theories of distributive justice upon which we might rely. I make the case for an ecumenical approach, arguing that the leading theories are likely to converge in identifying many cases as either just or unjust in the distributive sense. Using this approach, I acknowledge the many risks of distributive injustice that deployment of SRM might entail. Next, drawing from previous work with Morrow,[5] I provide an account of non-ideal distributive justice. On this account, some climate policy must be politically feasible, likely to be effective, and morally permissible in order to count as non-ideally just. This third condition requires some careful consideration in particular. Drawing again from my work with Morrow, I rely on a notion of permissibility employing both proportionality and comparative criteria. Roughly, to be permissible in this sense requires a policy to secure moral goods that are proportionate to the moral ills it might carry, as well as to compare well to other (feasible and effective) policy options when it comes to this ratio of goods to ills. I also argue here that climate policy is best approached as a matter of non-ideal justice, because some parties are failing to comply with their (ideal) obligations of justice (e.g., to cut emissions). Although SRM would not be deployed in an ideally just world, it is reasonable to ask whether in our world it might satisfy the three conditions just mentioned. Indeed, there is some reason to think that SRM might reduce risks of distributive injustice, potentially reducing harms that (for example) low emitters might otherwise experience. As it turns out, however, there is a tension between the political feasibility and moral permissibility of SRM.

For instance, unilateral or small-scale multilateral deployment of SRM could be politically feasible in virtue of sidestepping the need for extensive negotiation at the international level, but this would carry a substantial moral ill, namely the unfair exclusion of many interested parties from decision-making on matters that stand to affect them substantially. This does not show that SRM policies must fail to be non-ideally just, but it does suggest that it might be challenging for an SRM policy to satisfy all three conditions of non-ideal justice simultaneously. Finally, this second chapter addresses compensation for SRM-related injustices, offering an account that fits within a non-ideal framework.

Chapter three addresses procedural justice in decision-making about climate policy. I provide an account of procedural justice that is rooted in fairness. I adapt this from John Broome's theory of fairness regarding substantive goods. The idea is that fairness in decision-making holds when all parties with legitimate claims to contribute to some decision are allowed to do so in proportion to the strength of their respective claims. What I call "complete fairness" holds in some case when all parties with legitimate claims to the relevant decision-making have those claims fully satisfied in exact proportion to their strength. Absent complete fairness, decision-making can be fair (or, if you like, unfair) to varying degrees. What I call "incomplete fairness" holds in some case when some (but not all) claims to decision-making are at least partly satisfied. As a matter of non-ideal procedural justice, we should aim for as much (incomplete) fairness as is politically feasible. I then argue that signatories to the United Nations Framework Convention on Climate Change (UNFCCC) have legitimate claims to contribute to decision-making on global climate policy, because those parties have effectively promised one another the opportunity to make such contributions. This includes decisions about global climate policies that include SRM. Accordingly, unilateral or small-scale multilateral SRM would involve substantial procedural unfairness, because by definition it would exclude many parties that have legitimate claims to contribute to decisions about whether (and how) to deploy SRM. In the same chapter, I show that procedural justice has a pragmatic value in addition to its moral value. Specifically, in cases of (deep) uncertainty, those with legitimate claims are allowed to make their own decisions about how to proceed, rather like a patient who is afforded a choice among uncertain (and possibly dangerous) treatments.

In the fourth chapter, I turn to questions of virtue and vice in relation to SRM. At first glance, it might appear rather quaint to speak of virtues and vices in policy-making, but I argue that a lot rides on whether or not the agents of SRM are virtuous. To be clear, I neither defend nor rely upon a virtue-based moral theory in this chapter. Rather, I take it that a virtuous agent is (among other things) someone who is reliably responsive to moral reasons for action. For instance, a person with the virtue of justice can be counted upon to act in a just fashion in most cases. If, in some situation, the would-be agents of SRM lack relevant virtues, then we have reason to be concerned that SRM might not be deployed in a morally acceptable way. This would be so even if there are possible uses of SRM that would satisfy the demands of non-ideal justice, as those who deploy and maintain SRM would lack the traits (i.e., the relevant virtues) that make one reliably responsive to the moral reasons we have for favoring (non-ideal) justice. Worse yet, if the would-be agents of SRM carry certain relevant vices (e.g., greed), we have even greater cause for concern, for SRM might be used in self-serving ways that depart substantially from any use of SRM that might count as non-ideally just. I argue that virtuous climate engineers are possible, sketching the features they would possess. There are two reasons this is worthwhile, even for an approach (like my own) that is not committed to a virtue-ethical moral theory. First, by thinking about whether and how a virtuous agent would deploy SRM, we can clarify for ourselves what a morally appropriate use of SRM might look like—or, indeed, whether there is a morally appropriate use of it. Second, as already indicated, by asking whether such virtuous climate engineers are likely in our actual world, we can clarify the prospects (or lack thereof) for some morally appropriate use of SRM. One might conclude, for example, that although some use of SRM would be non-ideally just, there is virtually no chance that actual agents of SRM would use SRM in that fashion.

Chapter five addresses the question of genuine moral dilemmas, in the technical sense used by moral philosophers. In a genuine moral dilemma, all available courses of action entail moral wrongdoing, such that it is impossible to act in a permissible way. I develop the idea of a "pessimistic climate scenario," or a situation in which all available courses of action carry substantial moral problems, such as risks of unjust harm to innocent parties. I use the term "pessimistic" because (I assume) we are not yet committed to such a scenario holding in the future. However, if we continue to make little progress in working toward just responses to climate change, such scenarios are likely in the future. In some future scenarios that are likely to hold in a pessimistic future, it is tempting to think that SRM might be, in an all-things-considered moral sense, part of the best (or least) bad option for proceeding. Indeed, many proponents of researching SRM allow that other options (e.g., aggressive mitigation) are preferable, but they insist that we may need SRM in the future if we fail to pursue those options to a sufficient degree. For instance, despite the risks of injustice SRM carries, its potential to cool the climate quickly might allow us to manage even worse risks of injustice arising from past and ongoing emissions. However, even granting that SRM might be the best (or least bad) option in some future case does not automatically entail that deploying it would be permissible. Gardiner has argued that such cases could constitute genuine moral dilemmas, and so SRM could be morally impermissible even if it is better (morally) than any other option.[6] Whether such dilemmas actually occur is a controversial matter in moral theory, but conceiving of (some) pessimistic climate scenarios as such has several advantages. Doing so captures both the moral disvalue of many of SRM's likely impacts and the substantial moral failures (e.g., to reduce emissions) responsible for making SRM into the best option. However, I argue that viewing pessimistic scenarios as genuinely dilemmatic has serious downsides, contravening plausible moral principles (e.g., "ought implies can") and undermining moral action-guidance when it is most needed. Instead, I advocate an agent-regret account. I argue that, *if* SRM is indeed (part of) the morally best option for proceeding in a pessimistic scenario, then we ought to deploy it, but it is appropriate for this action to be accompanied by a genuine regret that seriously acknowledges both the moral disvalue associated with that use of SRM and the morally relevant context in which it is deployed.

In the concluding chapter, I bring together these various strands. What is needed is an approach to climate policy that takes justice seriously while also acknowledging the (largely political) constraints that have so far prevented anything close to ideal climate justice. Here I make the case that some uses of SRM could satisfy the conditions of non-ideal justice in some realistic, pessimistic scenarios. This is plausible if we take SRM to be part of a hybrid policy—for instance, one that includes commitments to politically feasible mitigation, adaptation, and perhaps CDR. In a pessimistic scenario, our ethical assessments of potential climate policies should be essentially comparative. Given that no feasible and effective option will be free of substantial moral problems, we must ask how the available, problematic options stack up in terms of their respective moral goods and ills. I discuss various ways of reducing risks of injustice associated with SRM. Nonetheless, despite these prospects for a non-ideally just use of SRM in the future, we must return to considerations of agent-regret. If it happens in the future that we find ourselves in a scenario in which SRM is, morally speaking, the thing to do, that would indicate a massive moral failure on the part of many of us. We should not lose sight of that important fact.

Notes

1. David Keith, "Geoengineering the Climate: History and Prospect," *Annual Review of Energy and the Environment* 25 (2000): 245–84.
2. Stephen M. Gardiner, *A Perfect Moral Storm: The Ethical Tragedy of Climate Change* (Oxford: Oxford University Press, 2011).
3. Eric Posner, "You Can Have Either Climate Justice or a Climate Treaty: Not Both," *Slate*, November 19, 2013, www.slate.com/articles/news_and_politics/view_from_chicago/2013/11/climate_justice_or_a_climate_treaty_you_can_t_have_both.html.
4. Clare Heyward and Dominic Roser, *Climate Justice in a Non-Ideal World* (Oxford: Oxford University Press, 2016); David R. Morrow and Toby Svoboda, "Geoengineering and Non-Ideal Theory," *Public Affairs Quarterly* 30, no. 1 (2016): 85–104.
5. Morrow and Svoboda, "Geoengineering and Non-Ideal Theory."
6. Stephen M. Gardiner, "Is 'Arming the Future' with Geoengineering Really the Lesser Evil? Some Doubts about the Ethics of Intentionally Manipulating the Climate System," in *Climate Ethics*, ed. Stephen M. Gardiner et al. (Oxford: Oxford University Press, 2010), 284–312.

1 Benefits

In considering how to respond to climate change, we should care a great deal about the outcomes various policies are likely to bring, given that different policies can diverge greatly in their geophysical and social impacts. We have various reasons to care about such outcomes, some of them prudential and some of them moral. I am interested in the latter. A natural way to think about policy outcomes is in terms of overall net harms and benefits. If we accurately aggregate the various harms and benefits for some policy, we can then determine the net benefit (or net harm, as the case may be) that policy would yield if pursued. If we do this for each of several policies, we can then compare those policies, ascertaining which one delivers the greatest net benefit (or the least net harm). It might seem obvious that we ought (morally) to prefer that policy which would result in the greatest net benefit. Roughly speaking, this thought is in line with a maximizing form of utilitarianism, or the family of normative ethical theories holding that morally right (and wrong) actions are determined solely in virtue of the outcomes those actions yield, with right actions taken to be those that maximize net benefit. As we shall see, however, this utilitarian thought would be too quick. While it is plausible to suppose that we have a *prima facie* moral reason to prefer the policy delivering the greatest net benefit, in some cases it is permissible (and sometimes even required) to pursue policies that do not maximize net benefit. There are two major reasons for this, which have to do with uncertainty and distribution.

To be clear, few if any advocates of researching SRM make their arguments in explicitly utilitarian terms. This is not surprising, given that the ethical commitments of arguments for SRM research tend to be left inchoate. However, many such arguments emphasize the potential for SRM to deliver benefits that might greatly outweigh its costs, making utilitarian characterizations at least plausible.[1] Indeed, proponents of researching SRM often contend that, despite some harmful impacts (e.g., due to precipitation change), SRM could reduce many of the harms of anthropogenic climate change by reducing or even halting planetary warming, such as by slowing the rate of sea-level rise due to melting ice sheets or reducing the frequency of extreme weather events.[2] We can use claims of this sort to construct a model utilitarian argument in favor of deploying SRM, holding that SRM would be permissible if the outcome of deploying it would be (on balance) more beneficial than that of any other available policy. Here no reference would be made to moral considerations that typically play a role in non-utilitarian theories, such as intention, justice, virtue, or other "moral constraints."[3] Instead, such an argument would treat the fact (if and when it is a fact) that SRM would produce the most beneficial outcome as a sufficient condition for its deployment to be permissible—and perhaps even obligatory. Like standard utilitarian arguments, the permissibility of the action or policy is taken to rest exclusively on the beneficial outcomes it would produce.

Although there is some plausibility to the claim that SRM could provide a comparatively attractive balance of harms and benefits in some cases, this line of reasoning faces a serious problem: the impacts of SRM, and climate change more generally, are uncertain. Climate model simulations provide only probabilities of certain impacts of SRM, such as regional changes in precipitation patterns. This uncertainty tends to be compounded when we move beyond purely geophysical impacts, such as when considering the implications of precipitation change for agricultural productivity, as well as the social consequences of any change in such productivity. So in comparing available climate policies in terms of their overall harms and benefits, we cannot simply assume that each policy will have a given set of outcomes. Instead, we need to take into account the probabilities of various impacts. There are methods for doing this, of course, such as utilizing expected utility theory. Unfortunately, there is often *deep* uncertainty (also known as ambiguity or Knightian uncertainty) regarding many of the impacts of SRM techniques, as well as climate change more generally. This means that we are uncertain about the probabilities themselves, often because different climate model simulations provide divergent estimates of these probabilities. As we shall see, this is a serious issue for utilitarian arguments regarding SRM, although deep uncertainty also poses problems for non-utilitarian approaches, including those reliant on some conception of distributive justice. A second problem for utilitarian arguments regarding SRM is that they overlook the question of distributive justice. Even assuming that some policy involving SRM would deliver the best overall balance of benefits to harms, those harms and benefits might be distributed in unjust ways. Although some consequentialist normative theories can take this into account (e.g., by using distributional weights in determining utilities), the relatively simple utilitarianism suggested by some arguments for SRM does not do so.

As a kind of toy model, it will be helpful to consider a simplified utilitarian argument in favor of SRM. I doubt that, on reflection, anyone would accept this argument, for reasons that we shall see. Moreover, I acknowledge that some forms of utilitarianism are quite sophisticated, providing nuanced ways to deal with uncertainty and distribution, a fact that is not represented by the model argument I am about to provide. Importantly, my goal is *not* to critique utilitarian (or, more broadly, consequentialist) assessments of climate policy. In fact, I will remain agnostic regarding the question of what normative ethical theory we ought to adopt. Rather, the purpose of this chapter is to raise concerns about a relatively simple, broadly utilitarian line of thinking that is suggested by some calls for researching SRM. The simple, model utilitarian argument is useful for helping us to see very clearly what problems arise for this type of justification. That argument runs as follows:

The Simple Utilitarian Argument for SRM Deployment

(1) In certain realistic future scenarios, there will be some SRM policy that would result in greater net benefit than any other climate policy.
(2) In deciding what climate policy to adopt, we ought to choose that policy which would result in the greatest net benefit.
(3) So in certain realistic future scenarios, it will be the case that we ought to adopt some SRM policy.

There are several things to notice about this argument. First, it does not predict that SRM will in fact prove to be part of the most beneficial policy. Rather, the first premise merely claims that this would be the case in some possible future scenarios. Moreover, the argument takes these to be "realistic" scenarios rather than outlandish ones, giving the argument some practical import. Of course, this premise might be false, and we probably lack sufficient justification for accepting it at the present time, given the early state of research on SRM. I turn to this premise in my discussion of uncertainty below. The second premise is what makes this argument a (simple) utilitarian one, for it directs us to consider *only* the resulting utility of the climate policies on offer, asserting that we ought to choose that climate policy resulting in the greatest net benefit. More specifically, this is in line with a "maximizing" utilitarian principle rather than a "satisficing" one. The latter would hold that we are morally required only to achieve some threshold of utility in our policy choices, such that adopting a sub-optimal climate policy could be permissible, provided that it is "good enough" in securing some specified level of aggregated net benefit. Conversely, the maximizing principle in the argument above directs us to choose that policy that would achieve the greatest net benefit. It would be easy enough to formulate a satisficing utilitarian argument for SRM, but that still would be subject to the problems under consideration. The problems with the Simple Utilitarian Argument are as follows.

First, because we are uncertain (in fact, deeply uncertain) about many of the outcomes of SRM, it is likely that we will not know whether some SRM policy will maximize utility (or satisfice with respect to the appropriate threshold) relative to other options.[4] Second, even if some SRM policy would be succeed in maximizing (or satisficing), there is no guarantee that it would deliver appropriate distributions of the relevant harms and benefits. The first problem is one of uncertainty, whereas the second is a problem of distributive justice. In what follows, I will refer to the Simple Utilitarian Argument and its maximizing principle, with the understanding that what I say would also generally hold for a satisficing version of that argument. It is important to realize that no advocate of researching or deploying SRM has explicitly endorsed the Simple Utilitarian Argument. As I have noted, the seemingly utilitarian overtones of calls for research on SRM—and, in some cases, even its near-term deployment[5]—are rather inchoate. Laying out this model argument is merely an initial attempt to become clear on what a utilitarian case for SRM might look like. Because of the problems just noted—which I examine in more detail below—we should search for a more sophisticated approach to assessing climate policies, whether that approach be utilitarian or non-utilitarian. In chapter two, I will defend an ecumenical approach to non-ideal distributive justice in assessing the distributional outcomes of various climate policies. In chapter three, I will defend the idea that procedurally fair decision-making provides an attractive method for dealing with (deep) uncertainty. First, however, we must take a closer look at the issues of uncertainty and distribution that SRM (and other climate policy options) unavoidably raise.

I should issue one more disclaimer. Any utilitarian ethical theory must specify what constitutes benefits and harms. Without some such conception, utilitarian principles of action, whether maximizing or satisficing, are not much use, as one cannot bring about beneficial outcomes without having some idea of what counts as such. Two prominent options are hedonism and preference-satisfaction theories. Hedonists, such as classical utilitarians, hold that benefits are constituted by pleasure, whereas harms are constituted by pain and the absence of pleasure. Adherents of preference-satisfaction theories maintain that benefits are constituted by satisfied preferences (or, more plausibly, *considered* preferences), whatever those happen to be, whereas harms are constituted by thwarted or unsatisfied preferences. There are important differences between these two types of theory, of course, but once again the problems of uncertainty and distributions are present in either case.

The Problem of Deep Uncertainty

Premise (1) of the Simple Utilitarian Argument claims that, in certain cases, there will be some SRM policy that, if adopted, would result in the greatest net benefit relative to other available climate policies. Unfortunately, it is difficult to know whether claims like this are true, because there is uncertainty regarding what the outcomes of any given climate policy will be. To be sure, we can model various aspects of given SRM policies, such as their impact on precipitation patterns and agricultural productivity, but these model simulations do not simply inform us what the outcomes will be, but instead yield probability estimates for different outcomes. Although, in principle, we can quantify the probabilities involved and say that some outcome is more or less probable than another (and by how much it is so), we cannot predict with certainty that some specific outcome will in fact occur should the policy in question be pursued. This is, of course, an elementary and obvious point about the nature of modeling studies, but it raises significant questions about how to make decisions when there is such uncertainty, as there is bound to be in decisions about climate policy.

Suppose we have several policy options in a given set of circumstances. If we knew all the outcomes of each policy and the net benefit or harm (or "overall value," as I will sometimes say) of each of those outcomes, then deciding what to do on utilitarian grounds would be a simple matter, at least in principle: we would need only to compare the guaranteed outcomes of the various policies on offer and choose that policy whose outcome had the greatest net benefit—or, in cases in which all available policies result in net harm, that policy which has the lowest net harm. But our policy choices for dealing with climate change are likely to depart quite a bit from this idealized case, given our uncertainty surrounding the outcomes of virtually any such policy. Because of this, deciding what to do about climate change cannot be as simple as choosing that policy which is guaranteed to have the greatest net benefit, simply because there are no such guarantees.

Fortunately, there are methods for dealing with uncertainty. Expected utility theory is the standard method. Suppose once again that we have several policy options in a given set of circumstances, but this time we are uncertain what the outcome of each policy would be. Instead, we know only that each policy carries a range of possible outcomes, some of which are more probable than others. This makes it unclear which policy would in fact deliver the greatest net benefit. Suppose, for example, that some possible outcomes of Policy A would be better than some possible outcomes of Policy B, but also that some possible outcomes of Policy B would be better than some possible outcomes of Policy A. To deal with this problem, expected utility theory directs us to take into account both the probability and the value of the possible outcomes of each policy. For Policy A, we assign a probability of occurrence to each of the possible outcomes, and we also assign a value (e.g., corresponding to the aggregated net harm or benefit) to each of these outcomes. For each possible outcome, we multiply the assigned probability by the assigned value. We then add these products (i.e., each outcome's probability multiplied by its value) together, and this gives us the so-called expected utility of Policy A. We do the same thing for Policy B, Policy C, and so on, until we have calculated the expected utility of all the policies on offer. (For our purposes, we can consider doing nothing to be a "policy," for that is an option which carries possible outcomes, each of which we can assign a probability and a value.) With this work done, we will be in a position to identify the policy with the greatest expected utility, and that is the policy we ought to adopt, at least according to standard expected utility theory.

A simple example to illustrate this might be helpful. Consider a choice between two options: A and B. Let us suppose that the possible outcomes

(O_n), as well as the attendant probabilities and values of those outcomes, for option A are as follows:

Option A: O_1 (.5, \$100), O_2 (.3, –\$50), O_3 (.2, \$5)

Option A has three possible outcomes: O_1, O_2, O_3. For each outcome, we have both a probability and a value. I am using monetary value here because it allows for simple and clear examples, although, strictly speaking, the utility or disutility of some outcome is distinct from its monetary value. Next, for each outcome we multiply the probability by the value, which gives us the following:

Option A: O_1 (\$50), O_2 (–\$15), O_3 (\$1)

Finally, to determine the expected value of option A itself, we merely sum these three products, which gives us \$36. This means that the expected (monetary) value of A is a gain of thirty-six dollars. We do the same thing for option B, which (let us suppose) has the following possible outcomes, each of which has the specified probability and value:

Option B: O_1 (.7, \$20), O_2 (.2, –\$30), O_3 (.1, –\$100)

We again multiply the probability and value of each outcome, which yields the following:

Option B: O_1 (\$14), O_2 (–\$6), O_3 (–\$10)

Now we sum these products to get –\$2, which means that the expected monetary value of option B is a loss of two dollars. Finally, we compare the expected monetary values of options A and B. In terms of expected monetary value, A clearly wins out: the expected value of B is a loss, albeit a relatively slight one, whereas the expected value of A is a gain. If we are committed to maximizing expected monetary value, and if these are the only choices available, then we ought to adopt option A. Of course, as I have already implied, in maximizing expected *utility* we do not seek to maximize expected monetary payouts as such. Rather, our aim is to maximize overall expected benefits. Due to the diminishing marginal utility of money, the same monetary payout will be more beneficial to some persons (e.g., the poor) than to others (e.g., the wealthy). To determine the expected utility of some option, we would need to rely on some utility function that ranks preferred outcomes, using this to assign utilities (as distinct from monetary values) to each possible outcome of that option. However, it is more intuitive to illustrate the relevant points using dollar payouts rather than units of well-being as such (e.g., so-called utils). Moreover, for most people, it is a safe assumption that greater monetary payouts will track greater increases in their utility, all else being equal, although the degree to which a given payout increases utility will depend on other factors, such as the current wealth of the recipient.

Expected utility theory is attractive in that it allows us to deal with uncertainty, but it is not itself an ethical theory. Instead, it is usually taken to be a theory pertaining to self-interest or prudence, providing tools for decision-makers to promote their own utility or that of some collective (e.g., citizens of some country). Nonetheless, expected utility theory can be incorporated into ethical theories.[6] This is perhaps obvious in the case of utilitarian ethical theories, but it requires some additional moves. Because utilitarians are typically interested in the well-being of all, the utility assigned to some outcome will be the aggregated utility of all (relevant) parties. This moves us toward a plausibly *ethical* use of expected utility theory rather than a prudential or self-interested use. Aggregating utilities is difficult, of course. There are practical limitations to how well they can be estimated, particularly if we take into account the well-being of future generations, as presumably we should, even if we weight them differently than the utility of present persons. In seeking to maximize expected utility aggregated across all (relevant) persons, we effectively adopt the perspective of a global social planner.[7] From this point of view, we determine which available climate policy has the greatest expected utility for all. This is calculated in the same way as before, with the difference that each assigned utility is aggregated across all relevant persons. We then adopt that policy with the greatest (aggregated) expected utility. This approach suggests another type of utilitarian argument for SRM deployment. Once again, this is a relatively simple argument, but laying it out is useful for identifying some of the issues that more sophisticated approaches—utilitarian or not—would need to address. This toy model argument runs as follows:

The Simple Expected Utility Argument for SRM Deployment

(4) In certain realistic future scenarios, there will be some SRM policy that has a greater expected (aggregated) utility than any other climate policy.
(5) In deciding what climate policy to adopt, we ought to choose that policy which has the greatest (aggregated) expected utility.
(6) So in certain realistic future scenarios, it would be the case that we ought to adopt some SRM policy.

Notice that, unlike the Simple Utilitarian Argument, this argument does not claim that some SRM policy *will* in fact deliver the greatest net benefit. In adopting the policy with the greatest expected utility, we might get unlucky, such as by realizing some low probability outcome that carries significantly less utility than the expected value of that policy. But we might get unlucky with *any* policy that carries a range of possible outcomes. Obviously, this argument has an important advantage over the Simple Utilitarian Argument, for it takes uncertainty into account. This is essential when it comes to thinking about climate policy. We need some way of evaluating policies not just for their potential harm-benefit ratios, but also for the probability of their delivering certain harm-benefit ratios. Yet the problem remains that climate policies are often *deeply* uncertain in terms of their outcomes. We are uncertain about the probabilities themselves, usually because different modeling studies provide divergent probability estimates for those outcomes. In cases of deep uncertainty, it is not clear what probabilities to assign to possible outcomes. Yet if we do not assign *some* probability to a given outcome, then we cannot calculate the expected utility of options that have this outcome as a possibility. This could make it unclear whether premise (4) is true, for it opens the possibility that some SRM policy would have the greatest expected utility on some probability estimates but not on others, thus making it unclear what we ought to do in such a case.

There are various options for dealing with deep uncertainty. We might start by appealing to a principle of strong or weak dominance:

> *Strong Dominance:* Given a choice between two options, A and B, if all the possible outcomes of A are better than all the possible outcomes of B, then we ought to prefer A.
>
> *Weak Dominance:* Given a choice between two options, A and B, if all the possible outcomes of A are at least as good as all the possible outcomes of B, and if there is at least one possible outcome of A that is better than at least one possible outcome of B, then we ought to prefer A.

Strong dominance entails that, for any possible pair-wise comparison of outcomes on A and B, A's outcome will be better than B's outcome. Weak dominance entails that, for any possible pair-wise comparison of outcomes on A and B, A's outcome will be at least as good as B's outcome, and there will be at least one such pair-wise comparison in which A's outcome is better than B's outcome. If some policy weakly or strongly dominates another, then on utilitarian grounds it seems reasonable to prefer the dominating policy. This is potentially useful in cases of deep uncertainty—as well as cases of strict ignorance, in which we lack *any* evidence about the probabilities of outcomes—because we need not assign probabilities to outcomes in order to evaluate their utility. Provided that we have identified all the possible outcomes and properly evaluated the utilities for both Policy A and Policy B—each of which is an admittedly tall order—we can determine whether one dominates the other. Let us suppose that we know Policy A does in fact dominate Policy B. We can then see that the former is to be preferred to the latter, even though we do not know the probabilities of the various outcomes of A and B. We are able to see this because whatever outcome ends up occurring, the outcome of A would be either better than or equivalent to the outcome of B, and this is the case because *all* the possible outcomes of A are better than or equivalent to *all* the possible outcomes of B. Another way to put this is that there is no possibility that B will turn out better than A in their respective impacts on overall net benefit, but there is a chance—or, in the case of strong dominance, a guarantee—that A will turn out better than B in this respect.

The drawback of such dominance-based approaches, of course, is that they work only under a constrained range of circumstances. Consider cases in which the possible outcomes of competing policies "overlap" in terms of their utilities, as when the best possible outcome of A is better than the best possible outcome of B, but the worst possible outcome of A is worse than the best possible outcome of B. In this case, neither option dominates the other. Without further information, we cannot safely hold that either one is to be preferred over the other. We might expect our choice among climate policies to involve many options among which a dominance relation fails to hold, given the wide ranges of possible outcomes indicated by modeling studies, not to mention the complexity of policy choices. Now perhaps the weak and strong dominance principles could be used as initial filters, disqualifying those options that are dominated (weakly or strongly) by others, but we would need some additional way to decide among those policies that remain.

One possibility is to adopt a maximin rule, according to which we ought to prefer that policy whose worst outcome is better than the worst outcomes of any other available policy. To illustrate this, suppose that the worst outcome on A is a very large reduction in overall well-being, whereas the worst outcome on B is a very small decrease in overall well-being. Clearly, the worst outcome on A is worse than the worst outcome on B. Accordingly, the maximin principle would have us adopt Policy B instead of A. Notice that this advice applies even if the best possible outcome on Policy B is substantially less good than the best possible outcome on Policy A. Because of this, we can think of maximin as a type of precautionary principle. There are many versions of the precautionary principle, but all advocate a "play it safe" attitude. With the maximin principle currently under consideration, the idea is that, in cases of deep uncertainty regarding the outcomes of various policies, we indeed ought to play it safe, adopting that policy whose worst possible outcome is better than the worst possible outcome of any other option. While it might not be obvious at first, a reliance on precaution can fit quite well with utilitarian thinking, whether of the maximizing or satisficing variety. The utilitarian's aim is not just to promote benefits but also to avoid harms. This opens the possibility that she will have good moral reason to prefer a policy that plays it safe, for although this approach might forego some possible benefits, it could also avoid substantial possible harms.

Initially, a reliance on maximin might seem very plausible. After all, if we don't know how probable the various outcomes are, would it not be wise to err on the side of caution in order to minimize potential harm? While this might be plausible under certain conditions, it is easy to envision scenarios in which following maximin seems to be quite bad advice. Suppose, for example, the possible outcomes of two options are as follows. Once again, I am using monetary value for convenience, but my comments will apply *(mutatis mutandis)* to utility proper.

Option A: O_1 ($1,000,000), O_2 (–$15)
Option B: O_1 ($–14), O_2 (–$13)

Since we do not know the probabilities of the possible outcomes, we cannot determine their expected monetary values. Nonetheless, we can evaluate the value of those outcomes themselves. Notice that there is neither a weak nor strong dominance relation between the two policies, so we cannot use a dominance-based approach even if we wished to do so. According to maximin, we should adopt option B, because its worst outcome (a loss of $14) is better (or less bad) than the worst outcome of option A (a loss of $15). This seems like poor guidance, however. Maximin would have us ignore the (admittedly unknown, but perhaps high) probability of receiving a million-dollar payout on option A. Many of us would be happy to risk losing an extra dollar or two in order to take the chance (perhaps low, perhaps high) of receiving an extra million. Intuitively, that stance appears to be rational, so we have reason to suspect that maximin should be disfavored as a guiding principle, at least in certain types of case. As with precautionary principles more generally, we might worry that maximin is too conservative, directing us to minimize potential harms at the expense of ignoring even very great potential benefits.

Other versions of the precautionary principle avoid being implausibly conservative, but often at the cost of being vague. For example, consider the version of the precautionary principle advocated in the Wingspread Statement:

> When an activity raises threats of harm to human health or the environment, precautionary measures should be taken even if some cause and effect relationships are not fully established scientifically. In this context the proponent of an activity, rather than the public, should bear the burden of proof. The process of applying the Precautionary Principle must be open, informed and democratic and must include potentially affected parties. It must also involve an examination of the full range of alternatives, including no action.[8]

This formulation of the precautionary principle is obviously not equivalent to maximin, but both share the idea that we should exercise caution when it comes to potentially harmful activities that are not well understood—for example, because we are deeply uncertain regarding their outcomes. To its credit, and unlike maximin, the Wingspread formulation does not straightforwardly entail that we should forego opportunities for great benefits in order to avoid small harms. Unfortunately, this is because the Wingspread formulation is so vague that it entails virtually nothing about how we should act, aside from directing us to examine all possible options and placing the burden of proof on proponents of potentially risky activities. Importantly, it does not tell us what suffices for meeting that burden of proof. This makes it very unclear how to act in cases of the relevant type. Moreover, by placing the burden of proof on those who advocate some activity, there is a worry that this precautionary principle will tend to privilege inaction, given that satisfying the burden of proof might be quite demanding. Yet privileging inaction could be a serious mistake in some cases, because doing nothing often carries its own risks of harm (or risks of missed benefits),[9] as with a "business-as-usual" stance toward climate change.

There are still other versions of the precautionary principle, and one might hope to find some formulation that avoids the problems noted so far. Minimally, we should want our precautionary principle to offer clear action-guidance in relevant cases, to avoid advocating clearly irrational courses of action (e.g., foregoing a chance at a great benefit in order to avoid a chance of a minimal harm), and to take account of the risks associated with doing nothing. Gardiner's "core precautionary principle," developed from John Rawls' *A Theory of Justice*, is a plausible contender. Rawls notes, "Clearly the maximin rule is not, in general, a suitable guide for choices under uncertainty. But it holds only in situations marked by certain special features."[10] Rawls identifies three such features: that we do not consider (e.g., because we do not know) the probabilities of different outcomes on the various options available to us; that we care "very little, if anything" for potential benefits beyond the minimum guaranteed by following maximin; and that the options disfavored by maximin include outcomes "that one can hardly accept" (e.g., because they are very harmful).[11] Given these three conditions, maximin can be appropriate in some cases, but it would *not* be appropriate to use maximin in our previous scenario, which was as follows:

Option A: O_1 ($1,000,000), O_2 (–$15)
Option B: O_1 ($–14), O_2 (–$13)

Whereas an unconstrained maximin rule would have us adopt option B, the Rawlsian maximin rule is plausibly taken not to apply in this case. While we do not know (and therefore do not consider) the probabilities of the various outcomes, we presumably do care a great deal about the potential benefits beyond the minimum ensured by maximin (a gain of one million dollars versus a loss of fourteen dollars, respectively), and it is presumably not the case that the option that would be disfavored by maximin (i.e., A) includes unacceptable outcomes. Because of this, the Rawlsian maximin rule does not tell us to prefer B over A, because the conditions of that rule's application are not satisfied.

Let us consider a case in which the Rawlsian conditions are plausibly taken to be satisfied. As before, I am using monetary value rather than utility for convenience.

Option A: O_1 ($30), O_2 ($20), O_3 (–$15), O_4 (–$2,000,000)
Option B: O_1 ($25), O_2 ($20), O_3 ($15), O_4 ($10)

If followed, maximin would have us choose Policy B, because its worst possible outcome (a gain of ten dollars) is better than the worst possible outcome of Policy A (a loss of two million dollars). Once again, we do not know (and hence do not consider) the probabilities of the various outcomes, so the first condition is met. The second condition is plausibly satisfied as well—few are likely to care very much about the difference between the minimum guaranteed by maximin (B's gain of ten dollars) and the larger gain that is possible on A (a gain of thirty dollars). Finally, the third condition also seems to be met, at least for individuals who are not very wealthy: Policy A includes possibilities of harmful outcomes that are unacceptable (e.g., losses up to two million dollars). Accordingly, maximin is applicable in this case, and it seems quite rational to follow its advice here.

Given this, perhaps Rawls' constrained maximin (and Gardiner's related "core precautionary principle") provides a helpful way to think about climate policy, including potential CE interventions. Roughly put, in cases of deep uncertainty, the Rawlsian maximin rule would direct us to prefer the available climate policy whose worst possible outcome is better than the worst possible outcome of *any* other available climate policy, provided that we do not care much about the additional benefits that would be possible on some other policy and that the other available climate policies carry the possibility of unacceptably harmful outcomes. Now let us imagine a situation involving deep uncertainty in which we know both what the possible outcomes are for each available policy and the appropriate utilities to assign to those various outcomes. In a case like this, the Rawlsian maximin rule— or a precautionary principle based on it—may not apply and thus will offer no guidance. Although the first Rawlsian condition is satisfied because we do not know the relevant probabilities, one or both of the other two conditions might not be satisfied. We have already seen this with the example discussed above:

Option A: O_1 ($1,000,000), O_2 (–$15)
Option B: O_1 ($–14), O_2 (–$13)

Here some of the Rawlsian conditions are not met, for we (presumably) care quite a bit about the potential gains on option A, which plausibly does not involve outcomes that are unacceptably harmful. Fortunately, it is intuitively obvious that we should favor Policy A, so the non-activation of maximin is not cause for concern here.

However, we may have some cases of deep uncertainty in which Rawls' conditions are not satisfied and in which it is not obvious how we ought to act, for example:

Option A: O_1 ($100,000), O_2 (−$1,500,000)
Option B: O_1 ($2,000,000), O_2 (−$2,000,000)

An unconstrained maximin rule would have us prefer A to B, but it is not clear whether the Rawlsian maximin rule would be applicable here. The first Rawlsian condition is satisfied insofar as we do not consider the outcomes' probabilities (again, because we do not know them), but the other two conditions might not be satisfied. Some people will care a lot about the difference between the minimum guaranteed by maximin (A's loss of 1.5 million dollars) and the potential benefit of B (a gain of two million dollars). Something similar might hold in some decisions about SRM.[12] Moreover, it is unclear whether we should count Option B's potential loss of two million dollars as an unacceptable harm in the relevant sense. To be sure, that would be a very great loss for almost anyone (although not for some very wealthy individuals or corporations), so it might seem obvious that the harm is unacceptable for almost anyone. Strictly speaking, however, this depends on our account of what makes an outcome unacceptably harmful. One puzzling feature is that, while a loss of two million dollars is intuitively an unacceptable harm, the same seems true for a loss of 1.5 million dollars. If so, then both options carry possibilities of unacceptable harm. If we must choose one or the other, then we cannot avoid the possibility of unacceptable harm no matter what we do. Yet if we followed maximin, we would be adopting an option (A) that carries a possibility of unacceptable harm, and then we might wonder why the possibility of unacceptable harm on Option B should count against it.[13]

The lesson here is that, while the Rawlsian maximin rule seems plausible, there may be cases in which it fails to be relevant. This is not a critique of the rule itself, but rather an acknowledgement that we are likely to be in need of additional tools in thinking about what to do in climate policy, particularly if we take a utilitarian approach to that question. Perhaps, as with weak or strong dominance, we should treat the Rawlsian maximin rule as one tool among others: in cases in which all three conditions are satisfied, we should adopt the policy whose worst outcome is better than the worst outcome of any alternative policy. But if those conditions do not all hold, and if there is no dominance relation among the relevant options, then how should we make decisions under deep uncertainty?

One option is suggested by Bayesianism, which (roughly put) allows us to turn decisions under deep uncertainty into decisions under mere uncertainty. We do this by assigning subjective "prior" probabilities (or, equivalently, "priors") to the various outcomes. This allows us to pursue expected utility maximization—by supplying probabilities for the outcomes, we can calculate the expected utility of each policy and then adopt the policy with the highest expected utility. This is potentially good news for a proponent of something like the Expected Utility Argument for SRM Deployment that I outlined above. The crucial question for this Bayesian approach is how the priors are to be assigned to the relevant outcomes when there is deep uncertainty (e.g., in the relevant scientific studies) regarding them. If there is nothing to guide this (subjective) choice of priors, we might worry that the assignment of probabilities will amount to little more than guesswork, an obviously dangerous approach given the high stakes of climate change and CE, particularly SRM. Fortunately, we need not resort to guesswork in this task, because deep uncertainty about probabilities does not entail pure ignorance about probabilities. Thanks to computer modeling studies, we have evidence regarding the potential impacts of various courses of action, including the impacts of different concentrations of atmospheric greenhouse gases (e.g., on global average surface temperature) and the impacts of certain uses of SRM (e.g., on regional precipitation patterns). We have deep uncertainty when two or more modeling studies provide divergent probability estimates for the same phenomenon. This divergence need not prevent us from adopting reasonable priors. First, considering all the relevant studies together might indicate some constrained range of possible outcomes, such as upper and lower bounds for some amount of precipitation change on a given SRM intervention. That is, all relevant studies might agree in finding that there is zero probability of some outcomes occurring. Provided that we lack reason to dismiss or discount the modeling studies in question, it is reasonable for all our priors to fall somewhere within this possible range. Of course, this range might be very large, so we need some further ways to narrow down our choices.

A second technique is to examine whether all the relevant modeling studies happen to concur on the probabilities of some of the outcomes. We might find such concurrence, for example, in the "tails" of the (perhaps otherwise divergent) probability ranges. The tails of a probability density function are the extreme possible outcomes (e.g., the maximum increase or decrease in regional precipitation) identified by the modeling study in question. These typically have a low probability that tapers off to zero. Now it might be that all the probability estimates from relevant studies yield approximately identical tails. In that case, our choice of priors should reflect this, with the probability we assign to the extreme outcomes reflecting the consensus among modeling studies on these particular outcomes. We might find similar consensus for other points on the probability estimates. If so, then it is reasonable to choose priors that reflect such consensus. In fact, an even stronger claim is plausible: unless we have good reason to dismiss or discount the relevant studies, it would be unreasonable to choose priors that are inconsistent with points of consensus among those studies. Of course, for this technique to be useful, we would require a fortuitous convergence of different models, which often does not occur. So this technique will be of limited use.

None of the foregoing helps us decide what prior to assign to outcomes that are admitted as possible by at least one modeling study, but to which different studies give divergent probabilities. Fortunately, there are methods for combining the results of different models into a single probability estimate. For example, Smith et al. use a Bayesian approach to combine the results of nine different climate models into a posterior distribution of future warming across various regions of the planet.[14] The details of such approaches are quite technical, and delving into them is beyond the scope of this chapter. What matters for our present purpose is that, given some appropriate way to aggregate the different results of two or more modeling studies, we can effectively transform decisions under deep uncertainty into decisions under mere uncertainty. We accomplish this by combining the various probability density functions of several studies into a single probability density function. This yields definite probabilities for the possible outcomes. If we can do this for each of the outcomes of a given option (or policy), we can determine the expected utility of that option. If we are able to do the same thing for the other available options, then we will be able to utilize the standard method of expected utility maximization, according to which we ought to adopt the option with the highest expected utility. This raises the question of how to weigh these divergent "viewpoints." Sometimes we may have reason to privilege some study over others (e.g., because it models some relevant factor not included in the other available studies). On other occasions, we might have no reason to privilege some studies over others, and so we should weight them equally.[15]

We should exercise caution in taking such an approach, of course. There is no guarantee that the resulting aggregated probability for some outcome is the "correct" probability. However, such an aggregated probability is a reasonable choice of prior, because it takes into account the evidence we have from all relevant studies. Moreover, in taking a Bayesian approach, these probabilities are always subject to updating in accordance with Bayes' Theorem. Without delving into technical details, this allows us to update our probabilities as we acquire new evidence from additional modeling work. Accordingly, even if our initial prior probability for some outcome is questionable (e.g., because it relies on limited modeling evidence), repeated applications of Bayes' Theorem on the basis of new evidence will allow us to refine our probability for the outcome in question. This process can be iterated, provided that we have new evidence that can be used for each subsequent application of the theorem. It is reasonable to hope that additional research on the potential outcomes of climate policies will help us reach fairly accurate probabilities for those outcomes. Accordingly, while the problems posed by deep uncertainty regarding outcomes are formidable, there are tools for dealing with them. This is good news for proponents of the Expected Utility Argument for SRM Deployment, as well as those sympathetic to utilitarian arguments for climate policy more generally. But other problems for such arguments remain to be addressed.

The Problem of Assigning Utilities

Perhaps a greater problem for utilitarian arguments for SRM—although not a problem unique to them—is that of assigning appropriate utilities to different outcomes. Setting aside the problem of uncertainty regarding what outcomes might occur upon adopting some policy, we still face "ethical uncertainty" regarding the value (or disvalue) of those possible outcomes. In order to determine what policy would have the greatest expected utility, where utility is understood in terms of aggregated net benefit, we need to know the net benefit or harm of all the relevant outcomes, for otherwise we would be unable to determine the expected utility of options that include those outcomes as possibilities. This requires some way of measuring and aggregating the pertinent harms and benefits of each possible outcome. We might think of this as cost-benefit analysis in a broad sense, for it attempts to take into account all the relevant goods and ills of different climate policies.[16] The goal is to determine the overall value or disvalue for each outcome, or its net harm or benefit on the assumptions of the Expected Utility Argument. Roughly put, we identify both the harms and the benefits of each outcome, weigh them up, and then determine the balance of benefits to harms. This balance will indicate whether the outcome's benefits outweigh its harms (or vice versa), as well as the extent to which that is the case. The result is the utility that we assign the outcome.

This is a difficult task. Suppose we have a reasonably well-described outcome for some SRM policy, having specified both the geophysical impacts and their social implications. How do we determine the net harm or benefit of the outcome? There are at least two major problems. First, there are competing theories of well-being—for example, hedonism and preference-satisfaction—which pick out different things as being beneficial or harmful. Adopting different theories of well-being might deliver different verdicts regarding what impacts are harmful or beneficial, as well as the degree to which they are harmful or beneficial. Second, we must aggregate these harms and benefits (however conceived) in order to determine the net benefit (or harm) of the outcome in question. Given the global impacts of most climate policies, particularly those involving SRM, this aggregative task will likely be daunting simply because it requires accounting for so much. Moreover, depending on the theory of well-being with which we work, some harms and benefits may be difficult to aggregate due to their (at least apparent) incommensurability, such as that between the value of a human life and the value of some commodity. Fortunately, both problems may be less difficult than they first appear, because many utilitarian approaches require only that we determine the comparative value of the available policies. The Expected Utility Argument for SRM, for example, claims only that we ought to choose the policy with the greatest expected utility. In determining this, we must assign utilities to various possible outcomes, but it might not be necessary to measure and aggregate the various harms and benefits in a *precise* fashion for each outcome. Reasonable estimates could be sufficient to indicate that some policy is clearly better or worse than some other in terms of their expected utility.

This tasking of weighing up harms and benefits might work well when the relevant values are purely monetary, or when they can be easily translated into monetary terms. Accordingly, we might choose some economic proxy of well-being, such as consumption. In principle, this has the advantages of rendering harms and benefits readily measurable and avoiding the problem of incommensurable values, as all harms and benefits are accounted for in terms of money. Of course, there is a concern that this approach risks missing what really matters. It is plausible to suppose some correlation between increased consumption and increased well-being up to a certain point. For instance, as any plausible theory of well-being will allow, extreme poverty tends to be an impediment to human well-being, as it inhibits the satisfaction of basic needs. As a general rule, increased consumption on the part of those previously living in poverty is reasonably expected to ameliorate harms and yield benefits to such persons. However, there are other cases in which consumption is less plausibly viewed as a reliable indicator of well-being.[17] Consider greatly increased consumption by someone whose previous levels of consumption were already quite high. It is not clear that such a person would receive much benefit from that additional consumption, whatever its magnitude. In fact, past a certain point, additional consumption might actually be harmful. This is not surprising if we reflect on what actually matters to us. Virtually no one values consumption (or any other economic good) for its own sake. Rather, we value consumption because of what it does for us (or allows us to do), such as promoting our happiness, satisfying our preferences, supporting functioning in accordance with our basic capabilities, and so on. At best, economic measures are proxies for well-being, and so we should exercise caution in using them to determine the harms and benefits of outcomes.

Instead of relying on economic metrics, we might turn our attention to harms and benefits that seem irreducible to monetary value, such as the inherent value of human lives. Climate change will result in many human casualties.[18] This is a very substantial harm, so an adequate accounting of the harms and benefits of climate policies must somehow take this harm into account. Broome argues that, in principle, the value of human lives can be adequately accounted for, but there are severe practical problems for carrying this out. He makes the assumption that the value of a human life is nothing more than the value of the good things within that life.[19] Admittedly, this is a controversial assumption, as some would argue that it fails to appreciate the very great value of human dignity possessed by all human persons. However, let us grant this assumption for the moment. Broome notes that we can use quantitative measures to evaluate some outcome in terms of human lives. We should not simply count how many lives are lost and saved in some outcome, for both the quantity and the quality of lives are plausibly relevant. The loss of one week full of boredom is not nearly as harmful as the loss of sixty years brimming with happiness. As Broome notes, a standard measure that takes these issues into account is the so-called disability-adjusted life year (or DALY), which measures both the loss of additional life years and any diminishment of the value of life years remaining to someone. Suppose that some individual has ten years of life remaining, but that adopting some climate policy now reduces his remaining life by four years *and* imposes some harm on him for the six years that now remain. This person has lost more than four DALYs, because the reduced value of her remaining six years is taken into account. The exact amount of DALYs lost will depend on the severity of that harm, as determined by some scale.

Such a quantitative measure of the value of human lives allows us (in principle) to account for them in determining the net benefit or harm of some outcome. But we also need our analysis to include more than just the value of lives, for other harms and benefits are plausibly relevant too, including those indicated by the changing value of commodities. For this to work, however, we need some constant measure of value, taking into account both the value of commodities and the value of lives (e.g., as measured by DALYs). Broome plausibly argues that we should not achieve this by converting the value of lives into monetary terms. To achieve such a conversion, we would need an exchange rate for the monetary value of a DALY. Broome notes that we could use a "willingness to pay" technique to determine this rate, with the monetary value of a DALY being the amount of money someone is willing to pay to achieve an extra DALY. However, because different people will be willing to pay different amounts for an extra DALY, this approach would have the implausible and ethically problematic implication that the lives of some persons are worth less than the lives of others. A wealthy person may be willing to pay much more than a poor person for an additional DALY, but only because the latter has little money that she can divert from meeting her basic needs. In a case like that, the approach just mentioned would entail that the lives of the rich count for much more than the lives of the poor in our analysis. Broome rightly rejects such an approach.

An alternative is to convert the value of things other than life into the units used to measure the value of lives (e.g., DALYs). Broome's reason for favoring this move (at least in principle) is that the value of a DALY is taken to be constant across different individuals, whereas the value of money is not. For one thing, a given amount of money typically has more value for poor persons and less value for rich persons. A billionaire receives little benefit from an extra ten thousand dollars, but the same amount might be extremely beneficial for a person living in poverty. Broome notes that we can use "distributional weights" to account for the fact that a given amount of money has more or less value for different individuals. However, we must begin with something that has constant value and then convert the value of other things into the units used for measuring this constant value. For example, because the value of a DALY is constant across individuals, we can convert monetary values for other things (e.g., commodities) into DALYs, provided that we use an appropriate distributional weighting to account for the differing value of a given amount of money across persons. In principle, a method like this can provide a quantitative measure of the value of both lives and commodities, allowing us to take both into account when determining the net benefit or harm of some possible outcome.

However, Broome notes that this approach faces difficult practical problems in the case of climate policy. We currently lack the information to perform an adequate accounting of the value of both lives and money, and so Broome advocates keeping our analyses of the value of lives and monetary values separate for the time being. To illustrate this, we might determine the net harm or benefit of each possible outcome of some policy solely in terms of DALYs, intentionally excluding monetary values. This may tell us something interesting, but the analysis will be incomplete, given that it ignores other kinds of value. Broome acknowledges that separating our consideration of the value of lives and the value of other matters "will leave a large hole in the cost-benefit analysis of climate change," but the (current) alternative of running a cost-benefit analysis that ignores both the value of lives and the differing value of money across persons is even more problematic. Moreover, Broome writes, "merging all values into one figure for benefits and another for costs may conceal more difficulties than it solves. It is probably better to keep the difficulties in the open, to make it clear that in the end the decision needs to rest on judgment rather than calculation."[20]

It is probably true that any decision about climate policy will involve a great deal of judgment, but the issues just noted raise serious challenges for utilitarian approaches to climate policy. To be fair, a similar problem also arises for non-utilitarian approaches that care about (among other things) the aggregated harms and benefits of policy outcomes. Yet this problem is especially poignant for broadly utilitarian approaches, considering how much importance they place on the (expected) value of such outcomes. In order to determine what available policy has the greatest expected utility, we need to assign approximately accurate utilities to the possible outcome of each policy, understood as the net harm or benefit of each outcome. But if this is to be useful to the utilitarian, the assessment must include (and properly weigh) all the benefits and harms countenanced by the operative utilitarian theory. As we have seen, utilitarian theories differ in their accounts of human well-being, but no plausible account thereof can deny that lives have value of their own (e.g., because of the pleasure or preference-satisfaction they contain). Indeed, we can argue quite plausibly that commodities are valuable solely because of the impact they have on the lives of persons, such as by bringing pleasure or satisfying preferences. Because of this, an analysis that ignores the value of lives runs a risk of missing something very important, at least from the point of view of a utilitarian with a plausible theory of well-being. Likewise, any plausible utilitarian theory will recognize that the value of money (and the value of commodities) differs across people. Again, the utilitarian will be interested in what some given quantity of money (or commodity) does for the person in question in promoting or hindering her well-being, and this will depend in part on how much money (or relevant commodity) that person already has. This is why it is natural for utilitarians to argue for transferring wealth from the rich to the poor (e.g., via effective charitable donations), as that wealth typically produces more utility in the hands of the poor.[21] Because of all this, the utility of some outcome as identified by a standard economic analysis will not be the utility that is of interest to the utilitarian. This is because most (but not all) such standard analyses make the "plainly false assumption that money has the same value everywhere,"[22] as well as excluding the value of lives. Accordingly, when the utility in question is understood according to utilitarian lights, we cannot use *standard* analyses to determine what policy has the highest expected utility. This is because such analyses usually do not account for features that are of crucial importance for utilitarian approaches.

Proponents of the Expected Utility Argument for SRM Deployment will need some method of assigning utilities to possible outcomes. Perhaps they could rely on a non-standard way of accounting for harms and benefits, such as that suggested by Broome, using appropriate distributional weights and converting all value into DALYs. As we have seen, however, there are serious practical problems for this approach, given that the relevant empirical work has not been done to a sufficient degree. At present, we are simply not in a position to determine (on a global scale) the net harm or benefit associated with specific, possible outcomes of climate policies. Future research might address this practical problem, providing the relevant data. Meanwhile, however, we might lack reason to accept the first premise of the Expected Utility Argument for SRM Deployment: "In certain realistic future scenarios, there will be some SRM policy that has a greater expected (aggregated) utility than any other climate policy." We cannot determine the expected utility of some climate policy if we do not know the utilities that should be assigned to the possible outcomes of that policy. Because of this, it is difficult to know whether the premise is true, raising serious doubts about whether accepting it is currently justified. Now there might be some realistic future cases in which the truth or falsity of this premise is evident despite substantial uncertainty of this ethical variety. For instance, as climate modeling and other areas of research progress, it *could* become clear that, in some future case, SRM provides the only way to avert some set of extremely harmful climatic impacts. In such a scenario, even if there is substantial ethical uncertainty regarding what utilities to assign to various outcomes, it might be obvious that SRM is *comparatively* better in terms of aggregated utility than the other options. Conversely, it might be obvious in some case that, despite ethical uncertainty, SRM is comparatively worse than some other option. However, although we cannot rule out this type of possibility *a priori*, it would be foolish to bank on it. Given cases in which it is not obvious how the policy options stack up in terms of their comparative utilities, we need some way to assign utilities to the various possible outcomes, and this requires some way of dealing with ethical uncertainty.

Another standard problem for assigning utilities concerns how to weight impacts on future well-being, particularly whether to discount future harms and benefits, and if so at what rate. Given a positive pure rate of time preference, the well-being of future generations is given less weight than that of the present simply because it occurs at a later time. There are two major questions here: should we discount future harms and benefits and, if so, at what rate? If we use a high discount rate for the future,[23] then harmful outcomes of some climate policy may count for very little if they occur in the further future. If we use a low discount rate[24] (or none at all), however, those very same outcomes could count for a lot, perhaps enough to give utilitarians decisive reasons for disfavoring the policy. Accordingly, how we answer the two questions just mentioned is very important. This is especially the case if we take a utilitarian approach to climate policy, because the discount rate on future well-being will drastically affect how future harms and benefits are weighted when estimating the net harm or benefit of a possible outcome.[25]

Philosophers tend to be critical of discounting future well-being. We might think that such discounting is ethically unjustified. Indeed, it seems arbitrary to give more or less weight to a given benefit or harm based solely on the time at which it occurs, as a pure rate of time preference would have us do. Nonetheless, we could still have other reasons to give priority of consideration to the present: "If people will be better off in the future, for instance, it is reasonable to be less concerned about their interests than those of the present generation, much as one might prioritise the less well off within a single generation."[26] In this case, however, we give more weight to the present than the future not because we accept some pure rate of time preference, but rather for a reason that is plausibly of ethical significance, namely the greater need of the present relative to the future. Now one might think that this just indicates that discounting future well-being may be justified after all, given the (presumed) greater well-being of future generations relative to that of the present. But this is not to adopt a *pure* rate of time preference. There is no necessity that the future be wealthier than the past. On the justification just suggested, we would deem discounting future well-being as appropriate if, and only if, the future will be better off than the present. Obviously, in cases in which the future will be worse off than the present, this justification would not hold, and so discounting future well-being would not be warranted. Yet in this same case, adopting a pure rate of time preference would require us to discount future well-being anyway. Consequently, a pure rate of time preference is vulnerable to the charge of arbitrariness, for it deems the mere fact that some harm or benefit occurs in the future as a sufficiently good "reason" for discounting that harm or benefit, regardless of the comparative well-being of present and future generations.

Economists sometimes argue that a pure rate of time preference should be used because it accurately reflects the undeniable fact that we do discount future well-being, both our own and that of future persons. This stance also suggests a way to answer the question of what discount rate to use for future well-being: we simply use the rate that our behavior implies. For example, a low rate of savings indicates a high discount rate on future well-being, revealing a time preference that gives a high priority to present benefits. If we think that economic models should work with the preferences that people actually have, then it makes sense to take the revealed discount rate as the one to use in our models.[27] Yet there are two serious concerns about this approach.

First, although it is true that human beings discount future well-being (their own and others'), that fact by itself does not provide reason to think the practice is ethically justified, much less that the revealed rate is justified. There is no doubt that some human beings have preferences that are self-serving or otherwise ethically questionable. This is particularly plausible in the context of climate change, which Gardiner likens to a "perfect moral storm" in which moral corruption is likely to occur.[28] The intergenerational aspect of that "storm" is salient to the question of discounting. Given that many of the worst harms of climate change will be felt by future generations rather than by ourselves, we may be less inclined to pay the costs of preventing or reducing these harms than we would be if they threatened us. Arguably, this merely reflects a selfish desire to enjoy the benefits of a carbon-intensive economy while shifting the burdens to other parties. Since many (but not all) of those other parties will be future persons, our behavior reveals preferences consistent with discounting future well-being. Yet this very behavior—as well as the preferences they reveal—have been subject to ethical critique.[29] That being the case, we cannot safely assume that the discount rate revealed by our behavior is ethically justified.

Now one might reply that the revealed discount rate is to be used because any other choice would be to impose a particular value judgment regarding what the discount rate ought to be. This was one criticism of the *Stern Review*, which used a low discount rate on future well-being.[30] Rather than relying on the discount rate already in play, Stern inserted a value judgment of his own into the analysis, an allegedly inappropriate move.[31] However, as Broome notes, value judgments are unavoidable when it comes to this issue.[32] Whether or not to discount future well-being (and at what rate) is an inherently ethical question. While it is true that Stern's low discount rate relies on a value judgment, so does the discount rate revealed by our current behavior. Collectively, humanity (or, more precisely, some segment thereof) has decided to value present benefits more than future benefits by a given amount. Whatever we think of this decision, it is surely a value judgment, because the decision involves giving different weights to different benefits (and harms). So we cannot defend the revealed discount rate on the basis that it avoids imposing values, because to adopt the revealed rate rather than some other is to make a value-laden choice. If the revealed discount rate is to be used, then some additional argument is needed to show this, just as would be the case if some other discount rate (or none at all) was to be used.

The second concern is that the debate around discounting the future tends to conflate two issues: discounting commodities and discounting well-being in the proper sense. Broome thinks that the former is justified while the latter might not be justified.[33] Commodities are those things that can be traded on markets and therefore have market values. Well-being in the proper sense involves the interests or flourishing of some entity. Different theories of well-being are available. Perhaps it consists of happiness, preference-satisfaction, functioning in accordance with one's basic capabilities, or something else. On any plausible theory, however, well-being will be distinct from the value of commodities. Therefore, even if we have good reason to discount the value of future commodities, we should be cautious about discounting future well-being. The conflation of these two objects of discounting might explain why economists tend to think discounting the future appropriate while philosophers tend to think the practice unjustified. Philosophers often see no good reason to adopt a pure rate of time preference for well-being proper, but economists tend to be interested in the value of commodities. Plausibly, there can be good reasons to adopt some discount rate for the latter. For instance, delaying the purchase of some commodity until a later date might allow one to pursue profitable investments in the interim, or future generations might be wealthier than the present one. In both cases, the commodity in question has less value in the future than it does in the present, and so it is reasonable to adopt some discount rate regarding it. Yet discounting future commodities in this way does not involve discounting the well-being of future persons. It is perfectly coherent to think that a future person's well-being is just as valuable as our own, but that a set quantity of some commodity is less valuable to that future person than it is to us at present. So perhaps we should discount the value of future commodities (at some rate yet to be determined), yet abstain from discounting future well-being proper.

These two concerns raise some difficulties for the Expected Utility Argument. How we address these concerns could make a substantial difference in terms of the utilities we assign to different possible outcomes, because these values are reached (in part) by aggregating utilities across time, thus making them susceptible to whether (and how) we discount future value. Any climate policy stands to have a large impact (harmful or beneficial) on both future commodities and future well-being. Because any choice of discount rate involves a value judgment, the proponent of that argument cannot avoid the issue by simply taking on the revealed discount rate. He needs to address the matter, deciding whether future value (regarding commodities, well-being, or both) is to be discounted, and if so at what rate. Any such choice is bound to be controversial, so those decisions will require defense. This creates the potential for quite wide-reaching ethical uncertainty regarding what utilities to assign to some possible outcomes of SRM (or other climate) policies. Harm in the future will count for much more if we use a low discount rate for future commodities and none for future well-being than if we use a high discount rate for both. Accordingly, where we stand on discounting matters quite a bit for expected utility calculations of policies having an impact into the future, but at present it is not clear precisely where we should stand on this issue.

Problems of Distribution

So far, we have been examining primarily epistemic problems with utilitarian arguments for SRM deployment, primarily the Expected Utility Argument. It is unclear that we would be justified in claiming, for example, that there will be some SRM policy that has the greatest expected utility in some set of circumstances, and this is due to both ethical uncertainty regarding utility assignments and deep uncertainty regarding the outcomes of different options. We now turn to an essentially ethical problem with the Expected Utility Argument. Consider the second premise of that argument: "In deciding what climate policy to adopt, we ought to choose that policy which has the greatest (aggregated) expected utility." It is this type of commitment that makes the argument in question recognizably utilitarian, for the claim is that the policy with the most beneficial aggregated outcome (here taken to be that with the greatest expected utility) ought to be preferred to the alternatives. But there are moral problems with this commitment. By focusing *solely* on the (aggregated) expected utility of some climate policy for some specified group, we overlook the distributions of harms and benefits across persons within that group. Intuitively, it matters how the harms and benefits of some climate policy are shared. If some segment of humanity is disproportionately harmed by our chosen climate policy, that is *prima facie* morally problematic even if that policy delivers the most beneficial aggregated outcome (relative to other available policies). This, of course, is in line with Rawls' well-known criticism of utilitarianism, namely that it does not "take seriously the distinction between persons."[34] Many will agree with Rawls that we should care not only about the aggregated harm or benefit of some policy, but also about how the disaggregated harms and benefits fall out among individual persons.

To be fair, although classical forms of utilitarianism might be open to Rawls' charge, utilitarians can accommodate this concern, such as by assigning distributional weights to benefits and harms across different populations. For instance, one might assign greater weight to benefits accruing to low emitters and lesser weight to benefits accruing to high emitters. It is important to remember that my aim in this chapter is not to critique utilitarian approaches to ethical assessment of climate policies. Rather, my aim is to point out that such approaches face some difficult challenges. I do not deny that some utilitarian approach might be able to overcome all these challenges in a satisfying way. The problem is that some justifications for SRM research (and potential deployment) implicitly rely on a relatively simple utilitarian claim, namely that the net benefits of SRM could outweigh its net harms. My view is not that we should reject utilitarianism in assessing climate policy, but rather that we should be careful when it comes to looking into SRM's utilitarian merits. It is well enough to point out that, in some cases, SRM could deliver net benefits that compare favorably to non-SRM policies, but this by itself is to remain silent on the ethically important question of how the disaggregated benefits and harms are shared. This question should be addressed. Of course, it can be addressed from various vantage points, including utilitarian ones.

I say that promoting the most beneficial aggregated outcome at the cost of disproportionately harming some parties seems morally *problematic* rather than morally *impermissible*. As we shall see in subsequent chapters, there may be cases in which producing uneven distributions of harms and benefits is permitted. But in some cases, uneven distributions obviously carry moral disvalue. Consider a climate policy that has the greatest expected utility but which provides substantial benefits to high emitters and substantial harms to low emitters. It is hardly obvious that this is the climate policy we ought to adopt, despite the fact that it secures the most beneficial outcome. Indeed, we seem to have a defeasible reason *not* to adopt this policy, given the disproportionate harm to low emitters. The obvious way to frame this concern is in terms of justice, particularly distributive justice. In selecting a climate policy, we should care about how the harms and benefits are distributed across affected persons—including, presumably, future persons. To this end, we need some principle (or principles) of distributive justice, which would specify the conditions under which distributions or harms and benefits are just or unjust. Of course, as with simple utilitarian approaches, we face issues of uncertainty here as well. I begin addressing this rather complicated task in chapter two, so I will not say more on that matter here.

To say that distributive justice matters is not to deny that we also have moral reason to favor policies that deliver greater net benefits than the alternatives. It would be a mistake to think that *only* distributive justice matters when it comes to climate policy. Consider a choice between two climate policies. Assume both will result in fully (and equally) just states of affairs, but that one policy will also bring a substantial net benefit to affected parties while the other will bring only a minor net benefit. If distributive justice was our only moral concern, then it would not matter (morally) which policy we choose, although we might have non-moral reasons to favor the one with the more beneficial outcome. That seems clearly mistaken, however. We *do* have moral reason (and not just prudential reason) to prefer the policy that secures both distributive justice and the greater net benefit. Even more strongly, we plausibly have moral obligations to promote both justice and general well-being—duties of justice and duties of beneficence, respectively—so we need to evaluate climate policies in terms of both these ends. While this is reasonable, it obviously complicates matters. However, it is not immediately obvious how our duties of justice and beneficence should be prioritized when they seem to conflict, nor is it immediately obvious whether and how we should trade off the relevant values when they do not happen to harmonize. We might think of distributive justice as a "moral constraint" on any potential use of SRM,[35] one that should be given priority over the goal of pursuing beneficence.

Like proponents of the utilitarian arguments considered above, those concerned about distributive justice are interested in policy outcomes. The difference is that the former chiefly evaluate outcomes for their net harm or benefit (or their expected net harm or benefit), whereas the latter chiefly evaluate outcomes for how harms and benefits are distributed across persons. But we also should care about the ways in which decisions about climate policy are made. The standard way to frame this concern is in terms of procedural justice. A principle of procedural justice specifies how decisions of a relevant kind ought to be made, including what parties ought to be included in decision-making processes. Although there is obviously a relation between how decisions are made and the outcomes those decisions are likely to yield, the question of procedural justice is distinct from questions regarding the moral value of policy outcomes. A procedurally just decision-making process *might* produce morally bad results, and morally good results might be produced by a decision-making process that fails to be procedurally just. Plausibly, procedural justice is ethically important in its own right, and we have moral reasons and obligations to promote it, but it is not the only thing that matters morally when it comes to climate policy. I will return to the question of procedural justice in chapter three.

Yet in taking justice seriously in these matters we face an immediate and obvious problem: climate justice seems politically infeasible, at least in the near term. Arguably, if powerful parties were seriously interested in promoting distributive justice, they would have accepted fairly aggressive mitigation policies in the past. So one might ask: what is the point of attempting to pursue climate justice in relation to potential SRM policies, given that such parties have signaled (rather clearly) their disinterest in justice? Although this (rhetorical) question might put things a bit too strongly, it suggests something important, namely that we need to address the issue of political feasibility. If an ethical assessment of global climate policy is to have any chance of making a social impact, it cannot ignore the issue of feasibility. Because of this, I will advance a non-ideal approach to justice, one which takes into account the social and political realities that make ideal climate justice unreachable at present.

Notes

1. Scott Barrett, "The Incredible Economics of Geoengineering," *Environmental and Resource Economics* 39, no. 1 (2008): 45–54; Paul J. Crutzen, "Albedo Enhancement by Stratospheric Sulfur Injections: A Contribution to Resolve a Policy Dilemma?," *Climatic Change* 77, nos. 3–4 (2006): 211–19.
2. Michael C. MacCracken, "On the Possible Use of Geoengineering to Moderate Specific Climate Change Impacts," *Environmental Research Letters* 4, no. 4 (2009): 045107.

3. David R. Morrow, "Starting a Flood to Stop a Fire: Some Moral Constraints on Solar Radiation Management," *Ethics, Policy & Environment* 17, no. 2 (2014): 123–138.
4. In the case of a satisficing approach, there is a possible exception to this, given that *all* of the possible outcomes of some SRM policy might happen to fall above (or below) the appropriate threshold. In that case, the probability of satisficing would be 1 (or 0, if all outcomes fall below the threshold).
5. Arctic Methane Emergency Group, "AMEG's Declaration," http://ameg.me/index.php.
6. Mark Colyvan, Damian Cox, and Katie Steele, "Modelling the Moral Dimension of Decisions," *Noûs* 44, no. 3 (2010): 503–29.
7. Richard Bradley and Katie Steele, "Making Climate Decisions," *Philosophy Compass* 10, no. 11 (2015): 799–810.
8. Wingspread Statement, "Precautionary Principle," 1998, www.sehn.org/wing.html.
9. Cass R. Sunstein, *Laws of Fear: Beyond the Precautionary Principle* (Cambridge: Cambridge University Press, 2005).
10. John Rawls, *A Theory of Justice, Revised Edition* (Cambridge: Harvard University Press, 1999), 133.
11. Ibid., 134.
12. David R. Morrow, "Ethical Aspects of the Mitigation Obstruction Argument against Climate Engineering Research," *Philosophical Transactions of the Royal Society of London A* 372, no. 2031 (2014).
13. On a similar point, Hartzell-Nichols argues that, because SRM carries its own threat of "catastrophe," researching (much less deploying) it is not justified as a precautionary measure against the threat of catastrophe posed by anthropogenic climate change. See Lauren Hartzell-Nichols, "Precaution and Solar Radiation Management," *Ethics, Policy & Environment* 15, no. 2 (2012): 158–71.
14. Richard L. Smith et al., "Bayesian Modeling of Uncertainty in Ensembles of Climate Models," *Journal of the American Statistical Association* 104, no. 485 (2009): 97–116.
15. Bradley and Steele, "Making Climate Decisions," 805.
16. John Broome, *Climate Matters: Ethics in a Warming World* (New York: W.W. Norton & Company, 2012), 101–4.
17. Michael J. Sandel, *What Money Can't Buy: The Moral Limits of Markets* (New York: Farrar, Straus and Giroux, 2012).
18. John Nolt, "Casualties as a Moral Measure of Climate Change," *Climatic Change* 130, no. 3 (2014): 347–358.
19. Broome, *Climate Matters: Ethics in a Warming World,* 161.
20. Ibid., 166.
21. Peter Singer, *The Most Good You Can Do: How Effective Altruism Is Changing Ideas about Living Ethically* (New Haven: Yale University Press, 2015).
22. Broome, *Climate Matters: Ethics in a Warming World,* 166; cf. Francis Dennig et al., "Inequality, Climate Impacts on the Future Poor, and Carbon Prices," *Proceedings of the National Academy of Sciences* 112, no. 52 (2015): 15827–32.
23. William D. Nordhaus, *A Question of Balance: Weighing the Options on Global Warming Policies* (New Haven: Yale University Press, 2008).
24. Nicholas Stern, *The Economics of Climate Change: The Stern Review* (Cambridge: Cambridge University Press, 2007).
25. J. Eric Bickel and Lee Lane, *An Analysis of Climate Engineering as a Response to Climate Change* (Frederiksberg: Copenhagen Consensus Center, 2009): 40.
26. Bradley and Steele, "Making Climate Decisions," 84.
27. Martin L. Weitzman, "A Review of the Stern Review on the Economics of Climate Change," *Journal of Economic Literature* 45, no. 3 (2007): 703–24.

28. Stephen M. Gardiner, "A Perfect Moral Storm: Climate Change, Intergenerational Ethics and the Problem of Moral Corruption," *Environmental Values* 15, no. 3 (2006): 397–413.
29. Simon Caney, "Cosmopolitan Justice, Responsibility, and Global Climate Change," *Leiden Journal of International Law* 18, no. 4 (2005): 747–75; Stephen M. Gardiner, "Ethics and Global Climate Change," *Ethics* 114, no. 3 (2004): 555–600; Henry Shue, "Global Environment and International Inequality," *International Affairs* 75, no. 3 (1999): 531–45.
30. Stern, *The Economics of Climate Change: The Stern Review.*
31. Nordhaus, *A Question of Balance: Weighing the Options on Global Warming Policies;* Weitzman, "A Review of the Stern Review on the Economics of Climate Change."
32. Broome, *Climate Matters: Ethics in a Warming World,* 111.
33. John Broome, "Discounting the Future," *Philosophy & Public Affairs* 23, no. 2 (1994): 128–56.
34. Rawls, *A Theory of Justice, Revised Edition,* 24.
35. Morrow, "Starting a Flood to Stop a Fire: Some Moral Constraints on Solar Radiation Management."

2 Distributions

We have seen that focusing exclusively on the aggregate, net harm or benefit of SRM overlooks the question of distributions. In deciding what to do about climate change, we should attend not just to the (expected) net benefit or harm of competing policy options, but also to the distributions of harms and benefits those options are likely to produce. The distribution consistent with maximizing (expected) net benefit *could* be unjust, and in that case we would have moral reason not to pursue the corresponding policy. Importantly, climate policies can be unjust to varying degrees. This is crucial to recognize. Under realistic future conditions, it may be that no climate policy can avoid some degree of distributive injustice. In a case like that, we might have an all-things-considered moral reason to adopt some policy that carries risks of distributive injustice, provided that it compares favorably to other available options in terms of justice (and other morally relevant considerations). Before making that case, however, we need to consider the risks of distributive injustice posed by SRM.

SRM appears to carry substantial risks of injustice. Its impacts are likely to be uneven, both spatially and temporally, and it threatens disproportionate harm to persons who bear little or no causal responsibility for anthropogenic climate change, notably the global poor and future generations. We may distinguish two varieties of distributive justice, that which holds (or fails to hold) at a given point in time and that which holds (or fails to hold) across time. I will refer to these, respectively, as intragenerational and intergenerational distributive justice. In this chapter, I am interested in what we may call *global* distributive justice. Because large-scale SRM, like many climate policies, will have substantial global impacts, there is a question of how its harms and benefits are likely to be distributed globally, and not only how its harms and benefits are likely to be distributed within individual states or regions. SRM's potential to disrupt regional precipitation patterns provides a good example of a risk of intragenerational injustice. Models suggest that SRM deployment could result in reduced average annual precipitation in some regions, raising concerns that this could lead to drought that would be harmful to various persons residing in affected regions.[1] Some regions likely to be most affected in terms of precipitation (e.g., southern Africa) have very low per capita emissions of greenhouse gases. Now suppose that some high-emitting country deployed SRM at the present time, and suppose this has harmful impacts on precipitation for low emitters. Intuitively, this harm would be unjust. If so, it would count as intragenerational injustice, because the agents and victims of that harm would be co-existent.

SRM's so-called "termination problem" provides a good example of a risk of intergenerational injustice. The most commonly discussed varieties of SRM, such as stratospheric injections of sulfate aerosols, require extensive upkeep in order to maintain a constant impact on atmospheric radiative forcing. Because injected aerosols would have a stratospheric life span of several years, a failure to maintain injections at some point in the future would result in rapid global warming.[2] This is because SRM would only "mask" the warming driven by atmospheric greenhouse gases. Once the injected aerosols dissipate, their counteractive cooling effect would also disappear, thereby allowing global average temperature to rise to what it would have been without SRM. Plausibly, such rapid warming could be very harmful, given the geophysical and social disturbances it might cause, including substantial economic damages.[3] Now suppose some generation decides to deploy SRM as a long-term policy, doing little to mitigate greenhouse gas emissions. This would put future generations at risk of SRM's termination (e.g., due to war or terrorism). Intuitively, this would be distributively unjust to future generations. In the course of addressing a problem for which future persons obviously bear no responsibility, the present generation would be burdening future generations with a risk of substantial harm. We can reasonably imagine future generations thinking that the past ought to have mitigated its emissions instead of saddling the future with the termination problem. Now it might be tempting to say that there is no problem here: as long as future generations maintain SRM they can avoid the harms of termination. This response would overlook the fact that SRM could involve harmful impacts (e.g., due to precipitation change) for future persons while it is in force. Ott has proposed that this forces future generations into a difficult dilemma: either continue using SRM "despite all the ills it causes," or stop using SRM and suffer the effects of termination.[4] Ott argues that it is morally wrong to impose such a dilemma on others. More specifically, we might say that doing so is a violation of intergenerational justice.

Theories of Justice

I have said that these two examples are "intuitively" unjust, but we need to move beyond intuition if our discussion of SRM is to be of much interest. What is required is a set of general considerations or principles regarding distributive justice. These principles help us to identify some states of affairs as distributively just or unjust, as well as to explain why it is so. Such principles are developed and defended in theories of justice, which (among other things) specify what distributive justice requires. We immediately have a problem, however: there are many theories of distributive justice, and it is a matter of reasonable debate which of them (if any) we should adopt. Thinking about distributive justice in climate policy might have been straightforward if there was consensus on this matter, but in truth it is not obvious what distributive justice exactly requires, and so there is legitimate debate among theorists.

One way to respond to this problem is to select some theory of justice and reject its competitors, using the selected theory to evaluate the distributive justice (or lack thereof) in various climate policies. This approach would have the drawback of requiring extensive elaboration and defense of the chosen theory, far more than is appropriate here. Further, the proceeding policy evaluation would stand or fall with the particular theory of justice on which it is based. This would make for a tenuous basis for that policy evaluation.

A better response to the problem is to see if there are points of convergence among different theories when it comes to SRM (and climate policy more generally). I have taken this approach in previous work with Keller, Goes, and Tuana.[5] If it is the case that two or more theories of justice converge in finding a given state of affairs just or unjust, and if our sole concern is to determine whether that state of affairs is just or unjust, then we need not attend to the theoretical differences among those theories. The idea here is that, although theories of justice obviously differ in their commitments in many important and interesting ways, those diverse commitments might yield the same (or at least similar) verdicts when it comes to many practical matters. Importantly, this can occur even if proponents of different theories provide different justifications for their verdicts. We may call this a "convergence hypothesis" regarding theories of distributive justice.[6] If this hypothesis holds true for SRM, then our analysis of SRM justice can proceed without settling on a single theory of justice. In what follows, I will sketch several prominent theories of distributive justice. After that, I will make the case that these theories are likely to converge in the case of SRM, at least on many important questions.

Desert-Based Theories

Some theorists argue that, when it comes to how benefits and burdens ought to be distributed, the chief criterion should be the degree to which various parties deserve those benefits and harms. This is a non-egalitarian type of theory. Desert can vary quite dramatically across parties, and so dramatically uneven distributions can be just, depending on the situation. Of course, in cases in which all parties happen to be equally deserving of some good, then desert-based theories will favor an equal distribution of that good, but such theories do not place any intrinsic value on equality as such. Desert can be understood in various ways. Two prominent conceptions appeal to contribution and effort, respectively.[7] On the former view, one's desert is proportional to her or his overall contribution to the social good. The more valuable one's contribution to the social good, the more deserving one is of receiving certain benefits and avoiding certain harms. On the latter view, one's desert is proportional to the effort she or he has put forth in the pursuit of certain activities. The greater one's effort, the more deserving one is of receiving certain benefits and avoiding certain harms. Both these conceptions of desert are readily applied to cases of monetary compensation (e.g., for employees). Arguably, they also can be salient in considering climate policies. We might ask whether the benefits and harms of some policy are likely to fall out in ways corresponding to the desert of affected parties—for example, whether the benefits are likely to go to those making the most valuable social contributions, whether the harms are likely to impact those inhibiting the social good, and so on. As with the other types of theories of distributive justice I will examine, I take no position on whether we should adopt a desert-based theory.

Egalitarian Theories

Many theories of distributive justice are egalitarian in some way, meaning that they take equality to be the chief consideration when it comes to how benefits and harms ought to be distributed. Such theories differ in how they answer the question, "Equality of what?" There are many things that we might identify as appropriate targets for equal distribution. One such view is that distributive justice requires there to be equality of well-being (or, equivalently, welfare) among persons. On such a view, a state of affairs is distributively just if and only if all relevant persons enjoy an equal level of well-being. However, while such equality might be a moral good in some sense, it is arguably problematic to take it as the aim of distributive justice. This is because someone's well-being can be enhanced or diminished due to one's own prudent or imprudent behavior. Intuitively, it is not unjust for someone to enjoy reduced benefits due to her own negligence, nor is it unjust for someone to enjoy greater benefits due to her own diligence. Desert-based theorists will tend to concur with this, and they might claim that equality should not be our (primary) concern when deciding how harms and benefits ought to be allocated. After all, the idea that negligence entitles one to fewer benefits than others is straightforwardly explained by claiming that such negligence makes one less deserving of the benefits in question. Yet one need not be a desert-based theorist in order to hold that distributive justice is sensitive to the voluntary choices of persons. Many egalitarians adopt some version of luck egalitarianism, which holds (roughly) that unequal distributions of harms and benefits are justified only insofar as such inequality affects persons due solely to their own responsible choices. On luck egalitarian views, the fact that someone enjoys a greater or lesser share of benefits than others can be just, provided that this is so entirely because of that person's own (say) diligence or negligence. Inequality that does not depend on one's own responsible choices—say, bad luck in the natural or social conditions into which one is born—is unjust on such views.

Equality of Opportunity

One form that luck egalitarianism may take is equality of opportunity for well-being (or welfare), according to which a state of affairs is distributively just if and only if such opportunity is distributed equally across all (relevant) persons. Arneson defends such a view of distributive justice: "For equal opportunity for welfare to obtain among a number of persons, each must face an array of options that is equivalent to every other person's in terms of the prospects for preference satisfaction it offers."[8] This plausibly counts as a luck egalitarian view, because Arneson admits that inequalities in opportunity are not unjust when resulting from "voluntary choice or differentially negligent behavior for which they are rightly deemed personally responsible."[9] On a theory like this, we evaluate a climate policy for its impact on opportunity for well-being. Some policies might do better than others in maintaining or promoting equality of opportunity, and others might introduce new (or exacerbate existing) cases of such inequality.

Equality of Resources

Another plausible case of luck egalitarianism is the view that distributive justice consists of an equal allotment of resources. Dworkin holds that justice requires an equal distribution of initial resources across persons, but he argues that justice does not prohibit future inequality of well-being owing to the (permissible) use of such resources.[10] From an initially equal distribution of resources, it is possible for a permissible series of transactions to result in a state of affairs in which some persons are better off than others. For example, of two persons with equal financial resources, one person might make wiser investments than the other, ending up with more beneficial returns. This is consistent with justice on a view like Dworkin's, given the equality of initial resources between the two. Consistent with a luck egalitarian commitment, the eventual inequality is not deemed unjust, because it results from the responsible choices of the individuals in question. If we use a theory like this to evaluate climate policy, then our interest will be in the potential for some policy to move us closer to or further from equality of resources among persons, as well as to maintain or undermine cases of such equality that might already exist.

Equality of Capabilities

We have seen two answers to the "equality of what" question: opportunity and resources. Another approach is to treat basic capabilities—specifically, capabilities to pursue valued functionings—as the benefits that are of greatest concern for a theory of distributive justice. These include (but, of course, are not limited to) capabilities to meet one's basic needs for nutrition and shelter, as well as to function as a member of a community. Sen argues that distributive justice requires that basic capabilities be equal across persons. He notes that other theories focus on "rights, liberties, opportunities, income, wealth, and the social basis of self-respect . . . rather than with what these good things *do* to human beings."[11] Arguably, capabilities are more important than these other goods, because the former constitute one's capacities for pursuing any valued functioning at all. Sen does not take equality of capabilities to be sufficient for distributive justice, but only necessary. Nonetheless, in principle we can use these ideas to evaluate different climate policies, such as for whether they promote or inhibit equality of capabilities across persons. However, because this does not constitute a full theory of distributive justice, we cannot safely conclude that some policy securing such equality is just, for it might violate some additional requirement of distributive justice.

Rawlsianism

Rawls holds that the appropriate principles of justice (including distributive justice) are those that would be chosen by rational and mutually disinterested deliberators behind a "veil of ignorance," with this preventing individuals from choosing principles that would *unfairly* benefit themselves or harm others. Behind this veil, "no one knows his place in society, his class position or social status, nor does anyone know his fortune in the distribution of natural assets and abilities, his intelligence, strength, and the like."[12] Rawls argues that deliberation behind the veil of ignorance would yield the following two principles, with the first having priority over the second.

> First: each person is to have an equal right to the most extensive scheme of equal basic liberties compatible with a similar scheme of liberties for others.
> Second: social and economic inequalities are to be arranged so that they are both (a) reasonably expected to be to everyone's advantage, and (b) attached to positions and offices open to all.[13]

Rawls' theory of justice is thus largely egalitarian, for it requires that basic liberties be distributed equally and that offices be open to all. Yet part (a) of the second principle—the so-called Difference Principle—allows for social and economic inequality when that is "reasonably expected" to be to the advantage of all, including the worst off. Because the first principle takes priority, however, inequality in basic liberties is not permitted even when that would be to the advantage of all. We may say that Rawlsian distributive justice requires that basic liberties be maximized to the extent that they can be enjoyed equally by all persons, that offices be open to all, and that social and economic inequality (if any) be such that it benefits all, including the worst off. If we were to rely on such a Rawlsian notion of distributive justice to evaluate some climate policy, then we should attend to how that policy is likely to impact basic liberties and socioeconomic equality. Some climate policy entailing social or economic inequality might be distributively just on this approach, but only if that inequality is to the advantage of all, including the worst off.

This has been a very brief rundown of some, but not all, prominent types of theories of distributive justice.[14] My aim has not been to say anything interesting about such theories themselves, but rather to indicate various things we could reasonably care about in examining the question of distributive justice with regard to climate policy. We might suspect that, given their differing theoretical commitments, these theories will tend to provide divergent verdicts on whether some climate policy is distributively just or unjust, as well as on the extent to which that policy is just or unjust. To the contrary, I will argue that many different theories are likely to converge on such verdicts, although each will provide a distinct justification for its evaluation of climate policy. This is easiest to see in the cases of distributive injustice. Let us reconsider the *prima facie* case of intragenerational injustice mentioned above. Suppose that SRM is deployed, reducing climatic risk but also resulting in drought, primarily in regions inhabited by persons with relatively low emissions. Plausibly, this is taken to be an injustice by all of the types of theory just surveyed, albeit for different reasons. I will take each type in turn, but first I should issue a disclaimer. The example being discussed is a possible outcome of some uses of SRM. I do not claim here that such drought would in fact occur if SRM was deployed, nor do I claim that, if it does occur, it would make that use of SRM unjust on the whole. Instead, as an illustration of how convergence might operate, I will argue only that this outcome of SRM, should it occur, is plausibly taken to be unjust by all the theories of justice mentioned above.

First, it is not reasonable to think that low emitters deserve the burdens imposed by drought, nor that persons in unaffected regions (e.g., high emitters) deserve the benefits SRM might secure for them, such as decreased vulnerability to the harms of anthropogenic climate change. This goes for both contribution and effort conceptions of desert. Plausibly, there is not (on average) any relevant difference in the effort (e.g., in one's work) put forth by low emitters and high emitters that would justify placing a greater burden on the former. Nor is there reason to suppose that low emitters make (on average) less of a contribution to the social good than high emitters. Actually, the reverse might be true: due to the harm caused by emissions, high emitters arguably make less of a contribution to the social good than low emitters, all else being equal. If anything, then, a desert-based theory of justice that conceives desert in terms of contribution could entail that greater burdens should be borne by high emitters. At any rate, such a theory gives us reason to deem SRM-induced drought in low-emitting regions to be unjust, given that many of those affected by it would not plausibly deserve to bear the associated harms.

The same verdict is likely to be reached by egalitarian theorists, although not due to considerations of desert. An outcome that disproportionately harms some (e.g., low emitters) while benefiting others (e.g., high emitters) would exacerbate inequality of well-being between those two groups. As we have seen, however, pure equality of well-being is not a plausible requirement of distributive justice. Luck egalitarians, however, are likely to find our example unjust. Harmful, disproportionate drought plausibly undermines both equality of opportunity for welfare and equality of resources. Freshwater is a resource, and access to it is obviously a basic human need. If someone's access to freshwater is diminished, that will make it more difficult for him to meet a basic need. He will need to expend more time, energy, or money to meet that need, or he will simply suffer the harm of having that need met to a less than optimal degree. Accordingly, both his resources and his opportunity for welfare have been damaged, because he now has less of a resource than he once did, which makes it more difficult for him to maintain his previous level of welfare. Because, in the case envisioned, SRM disproportionately has these impacts on low emitters, it therefore introduces (or increases) inequality of resources and inequality in opportunity for welfare. On views like Dworkin's and Arneson's, the problem with SRM-induced drought is not merely that it exacerbates inequality, but that it does so in ways not attributable to the responsible choices of those who are detrimentally impacted by it. Assuming that, by and large, the victims in question are not also voluntary agents of SRM, those impacted by drought have their resources and opportunity for welfare damaged through no fault of their own. So this exacerbation of inequality would not be justified on standard luck egalitarian views, and so from the vantage of such views we have reason to regard SRM-induced drought as distributively unjust.

The case is even more straightforward for a view that takes equality of capabilities to be a necessary aim of distributive justice. Any plausible list of capabilities will include the capability to meet basic health and nutritional needs. If someone's access to freshwater is compromised, the capability just mentioned will also be compromised. If, on the whole, SRM compromises this capability of low emitters without impacting the capabilities of high emitters, then a proponent of something like Sen's approach would have reason to judge that use of SRM to be unjust.

Finally, a Rawlsian has at least one reason for concern about our envisioned case. First, we can reasonably expect harmful drought to threaten the enjoyment of basic liberties on the part of affected persons. It is difficult to exercise one's right to free expression, for example, if drought forces one to shift substantial attention to meeting the basic needs of oneself and one's dependents. Arguably, this would violate Rawls' first principle, namely that "each person is to have an equal right to the most extensive scheme of equal basic liberties compatible with a similar scheme of liberties for others." Now one might object that this principle actually is not violated in the envisioned case. Unlike (say) an oppressive government that affords certain rights to some while denying them to others, deploying SRM does not involve *denying* basic liberties to some persons, even if it the outcomes of SRM make it more difficult to exercise their liberties. Even if this objection is correct, however, our case of SRM-induced drought in low-emitting regions seems to violate the Difference Principle, which requires that social and economic inequality be "reasonably expected to be to everyone's advantage." Because such drought disproportionately affects some persons, we can reasonably expect social and economic (e.g., due to negative impacts on agricultural productivity) inequality to be exacerbated. This is not *reasonably* expected to benefit all, because many of those affected by SRM-induced drought are likely to be harmed (on balance). So there are Rawlsian reasons for taking such drought to be a case of distributive injustice.[15]

Despite interesting and important theoretical differences, these theories converge in providing the same assessment of a particular kind of case. Here we need not delve into the complicated work of determining which theory we ought to adopt. Provided that one of these theories of justice is correct, and provided that we have good reason to expect convergence among them on the question of whether some policy is just or unjust, then we can rely on any one of these theories in order to determine whether that policy is just or unjust. This is not to deny that there may be cases of divergence among different theories of justice. No doubt there are such cases. When we have reason to suspect this we cannot rely on the approach just mentioned. Yet a convergence hypothesis of distributive justice for climate policy is generally very plausible. Many of the harms of interest to different types of theorist tend to reinforce one another: if someone's basic capabilities are diminished, her opportunity for welfare is likely to be as well; loss of resources will often lead to social and economic equality; if someone is harmed otherwise than through his own responsible choices, it is likely that he does not deserve to bear the associated burdens; and so on. The same holds *(mutatis mutandis)* for the benefits that are of interest to competing theories of distributive justice: enhancement in one's basic capabilities is likely to improve one's opportunity for welfare; the fruits of responsible choices will often overlap with deserved benefits; a transfer of resources to those who lack them is reasonably expected to diminish social and economic inequality; and so on. To be clear, I claim no necessary connection among these various harms (nor among these various benefits), but they are likely to go together, and this makes it reasonable to adopt the convergence hypothesis regarding theories of justice. I will rely on this hypothesis in what follows.

We need to address some further issues before continuing. First, we should distinguish between axiological and deontic uses of distributive justice.[16] Respectively, these uses involve treating justice as a value and treating justice as a duty. On axiological uses, we are interested in the moral value or disvalue of states of affairs—for example, whether it is a good or bad thing that harms and benefits are distributed in some fashion. On deontic uses, we are interested in the moral status of actions and policy choices—for example, whether they are impermissible, permissible, obligatory, or supererogatory. These uses of justice are not unrelated. Whether some action or policy is deontically just will depend in part on whether the state of affairs that action or policy produces or maintains is axiologically just. Yet these uses are nonetheless distinct. Injustice in the axiological sense is (morally) a bad thing, even if no one acted unjustly in the deontic sense in bringing about that state of affairs. Imagine a climate policy that involves a distribution of benefits and harms that is not in accordance with the distribution specified by the appropriate theory of distributive justice (whatever it is). Axiologically, this would be a case of distributive injustice. Deontically, however, this policy might not be unjust. This is because a given policy might be permissible (in light of our duties of justice) in some cases but impermissible in other cases. Plausibly, the question of permissibility hinges partly on how some policy compares to other available options, which of course will vary depending on context. For instance, in some unfortunate situation, it may be practically impossible to avoid some degree of axiological injustice. Given such a case, it is at least reasonable to think that it would be permissible to adopt a policy carrying some axiological injustice. As we will see, such conditions are likely to hold in the future. Because of this, simply pointing out that some climate policy carries risks of unjust distributions is not sufficient to show that this policy ought not to be adopted. As I will argue in chapter six, we also need to compare available policies to one another, assessing their respective prospects for (axiological) justice or injustice. I will make the case there that a policy can be deontically just even if it produces axiological injustice. From the point of view of distributive justice, some imperfect policy might be the best (or least bad) option available, and so we could have moral reason to pursue it despite its flaws.

Non-Ideal Theory

When it comes to climate policy-making, any appeal to principles of distributive justice faces the following challenge. Suppose that there are some possible climate policies that would satisfy the deontic requirements of justice, but that implementing any of these policies—for instance, a rapid transition to renewable energy on a global scale, with the transition in less developed countries paid for by more developed countries—would be infeasible due to political factors. Suppose further that there are some politically feasible climate policies, but that none of these would satisfy the requirements of justice. This plausibly describes our own world.[17] What use is a justice-based evaluation of climate policy in such a context? Proponents of climate justice risk wasting resources if they advocate measures whose practical prospects are very dim. So perhaps such proponents should instead advocate the best (or the least bad) policies that are feasible. Yet this second approach would seem to be a departure from considerations of justice, for one would then be advocating policies that fail to satisfy principles of justice. The challenge, then, is to specify how considerations of distributive justice can be practically useful in our present context.

A very promising answer to this challenge is to investigate climate justice under the auspices of "non-ideal theory." As opposed to ideal theory, which investigates what justice requires assuming full compliance with our duties of justice, non-ideal theory concerns what justice requires in cases of only partial compliance with such duties. It is clear that humanity's failure to cut greenhouse gas emissions is a case of non-compliance with our duties of (climate) justice, and so it is reasonable to take a non-ideal approach to climate justice in general.[18] Following Rawls, this involves searching for policy options that are politically feasible, likely to be effective in satisfying some non-ideal threshold of justice, and (in a sense specified below) morally permissible.[19] Unlike ideal theory, non-ideal theory takes into account contingent facts about the world, such as the prejudices of individuals or the intractability of governments. Attending to such facts might reveal that some ideally just policy has no chance of being taken up, and advocating for it might be ineffectual or even counterproductive (e.g., due to backlash from others). Those concerned with justice might then look for some politically feasible policy that is effective in securing a sub-optimal degree of justice while also satisfying certain constraints of moral permissibility. As we shall see, this non-ideal approach potentially gives SRM—and other imperfect climate policies—more traction than it would otherwise have. In an ideal world, presumably we would have mitigated our emissions long ago, such that CE would not be on the table. But in the actual, non-ideal world, we have at best only partially complied with our duty to mitigate our emissions, and the prospects for full compliance in the near future are dim. Accordingly, we need to consider what justice requires given the fact of partial compliance, and this is the province of non-ideal theory. Most of what I say in this section concerns the deontic use of justice, for our question is what policies are prohibited or allowed given the fact of collective non-compliance with our (ideal) duties of justice. In chapter six, however, I will make the case that even a non-ideal approach to climate justice should be comparative, which involves taking into account the axiological justice or injustice of available policies.

Before doing so, I wish to address a general objection to relying on non-ideal theory in the case of climate policy. Moellendorf notes that, at least on a Rawlsian approach, we require an ideal theory of justice before we can craft a non-ideal theory of justice, with the former providing a "compass" for the latter. Because climate change "will not wait until we have achieved agreement about matters of ideal justice," Moellendorf argues that there is a "lack of practicality of the project of developing a non-ideal account of justice for climate change."[20] On this argument, there is ethical uncertainty regarding what ideal justice requires. On the assumption that ideal theory provides needed guidance for non-ideal theory, this creates ethical uncertainty regarding non-ideal justice as well. With time, the global community might be able to sort through this uncertainty, coming to agreement on both what ideal justice demands and what this means for non-ideal justice. In the case of climate change, however, we lack sufficient time for this. Hence, non-ideal theory is not a practical option, and we should seek other approaches to thinking about the ethics of climate policy.

This is an overly pessimistic stance, however. Although I agree with Moellendorf that non-ideal theory requires a compass of sorts, that compass need not be a fully worked-out theory of ideal justice, nor must it be a theory that enjoys universal assent within the global community. As we have seen, given our convergence hypothesis for theories of distributive justice, it is reasonable to expect a good deal of overlap among competitors for the best ideal theory of justice, especially when it comes to identifying cases of injustice. This is because some injustices are simply obvious, and no credible theory of ideal justice will pronounce otherwise. Despite ethical uncertainty, we nonetheless have some idea of what ideal justice involves, and this may be good enough for crafting a working account of non-ideal climate justice. For instance, whatever ideal justice requires with regard to climate change, we can safely assume that it precludes policies saddling low emitters with heavy burdens while granting free rein to high emitters.

As a distinct response to the current objection, I also favor what is called "clinical theory," or the type of non-ideal theory that is concerned with averting and limiting existing or impending injustices.[21] This is a pressing question in the case of climate change, as there are many emissions-driven injustices that we have ethical reason to reduce or avert. Other types of non-ideal theory are concerned with *transitioning* to an ideally just society, seeking policies and institutions that will help a society achieve that long-term goal. Offering a (complete) non-ideal theory of justice in this transitional sense is an ambitious task. It may be, as Moellendorf says, that we lack sufficient time for transitional non-ideal theory to be practical in the case of climate policy. But this concern is less obviously applicable to a clinical-theoretic approach to climate justice. My question is the relatively near-term one of how climate injustice can be reduced, not the relatively long-term question of how we might transition to an ideally just society. Clinical theory is plausibly taken to be practical with regard to climate policy, both because many of the impending injustices of climate change are obvious and because (in one sense) clinical theory is less ambitious than transitional forms of non-ideal justice. I turn now to presenting an account of non-ideal justice in the clinical-theoretic sense. By the end of this book, the reader will be able to judge whether, as I think, this account and my application of it avoid Moellendorf's objection.

In previous work with David Morrow, we investigated the case for both CDR and SRM from the vantage of non-ideal theory.[22] Specifically, we employed clinical theory in the sense specified above. A clinical-theoretic approach is appropriate in the context of assessing CE policies, particularly those involving SRM. Very few would take SRM to be desirable in its own right, but it has been argued that SRM could play a role in reducing climatic harm or buying time for emissions mitigation.[23] As Morrow and I maintained, such arguments are plausibly understood as implicit appeals to clinical theory. In an ideally just world, humanity would have pursued substantial mitigation of emissions some time ago, in which case there would be little reason to consider SRM. Given our collective failure to pursue sufficient mitigation, we now face the prospect of substantial injustices associated with climate change. How we should deal with such injustices is a pressing question, and it is precisely the type of question clinical theory seeks to answer. This foregrounds a reasonable question: granting that SRM would violate principles of ideal distributive justice, might some SRM policy manage climate injustice in such a way that it is non-ideally just in a clinical sense?

We argued in that paper that clinical theory provides strong support for CDR methods of CE, but relatively weak support for SRM methods. The difficulty for SRM is that there is likely to be a tension between political feasibility and moral permissibility. Before giving the argument for this, we must examine what moral permissibility means here. Compared to the other two conditions for non-ideal justice (i.e., political feasibility and effectiveness), it is less clear what the moral permissibility condition requires. Rawls' idea seems to be, reasonably, that there are limits to what we may do in pursuing non-ideal justice. For instance, it would not be justified to kill innocent people in order to redistribute their organs to those in need, even were doing so politically feasible and likely to result in a substantially more just distribution of organs. However, a central commitment of non-ideal theorists is that certain actions that would not be justified in an ideal world can be justified in a non-ideal world, so the moral permissibility condition cannot be equivalent to whatever is justified under ideal conditions. The question, then, is what determines moral permissibility under non-ideal conditions. A helpful answer to this question must specify how the same policy can be prohibited under ideal circumstances but allowed under non-ideal ones.

Morrow and I provided the following account of the moral permissibility condition. In order to be morally permissible in non-ideal circumstances, a policy must satisfy both a proportionality criterion and a comparative criterion. On the former criterion, although a non-ideally just policy may carry various types of moral ill, such ills must be proportionate to the good that is achieved. This prohibits policies that would impose substantial ills in order to achieve relatively modest gains in alleviating injustice. On the latter criterion, a non-ideally just policy must compare favorably to other politically feasible options, in terms of both the injustice alleviated and the ills imposed by such policies. This prohibits adopting policies that are, from the point of view of justice, inferior to other politically feasible options. Given these two criteria, we have a notion of moral permissibility that is appropriately flexible, allowing us to take into account changing conditions when it comes to determining whether some policy is (non-ideally) just. At the same time, these criteria put serious constraints on what policies may be pursued under non-ideal circumstances.

SRM has the potential to reduce various risks of harm associated with climate change. Many of these risks are likely to be borne (unjustly) by those who bear little or no causal responsibility for climate change, such as the global poor or future generations. It is conceivable that SRM could be used in ways that would alleviate such risks of unjust harm, but whether such uses of SRM would be favored by clinical theory would also depend on whether those uses would be both politically feasible and morally permissible in the sense just described. Certain uses of SRM seem to be politically feasible, given the low expected cost of stratospheric aerosol injections,[24] the availability of the required technology,[25] and the possibility of unilateral deployment.[26] Arguably, aggressive mitigation policies have been politically infeasible because they tend to be expensive, require a massive deployment of the technology needed for alternative energy, and face difficult collective action problems in achieving cooperation among various parties. If some SRM policy could largely avoid these difficulties, then such a policy is plausibly viewed as politically feasible. This is not to say that SRM would be feasible for all states. As we noted, "Economically and militarily powerful states would presumably not permit less powerful ones (e.g., a coalition of small-island states) to deploy SRM against the powerful states' wishes. However, unilateral or minilateral SRM does, for better or worse, seem politically feasible for those (coalitions of) states that are sufficiently powerful to be able to ignore or discount the interests of other states."[27]

We can also conceive of realistic uses of SRM that might satisfy the moral permissibility condition. In order to meet this condition, some SRM policy would need to carry moral ills that are not disproportionate to the good achieved, and its ratio of ills imposed to goods achieved would need to compare favorably to the ratios of other politically feasible options. Under certain conditions, some use of SRM might succeed on these fronts, perhaps by slowing the rate of climate change and thus buying time for a large-scale transition to renewable energy, for vulnerable parties to adapt to changing environmental conditions, or both. These goods could be of a sufficiently great magnitude to justify the burdens imposed by SRM, such as drought in some regions. As Morrow and I noted, "While causing droughts would be wrong in ideal circumstances, doing so in this case might be permissible if the slower pace of adaptation relieved enough suffering," assuming that there are no other politically feasible options that would do better than SRM in terms of the ratio of goods achieved to ills imposed.[28] SRM could fare well on the comparative criterion due to its potential to cool the planet quickly, whereas other responses (e.g., mitigation and adaptation) tend to be slower. Of course, some uses of SRM would be clearly impermissible even under non-ideal circumstances, such as a case in which SRM achieved relatively modest benefits in more developed countries at the cost of imposing substantial harms in less developed countries. In a case like that, some particular use of SRM would fail to satisfy the proportionality criterion. It would also fail to satisfy the comparative criterion, provided that there are better, politically feasible options that may or may not involve SRM.

So far, we have seen reasons to believe that *some* particular uses of SRM could satisfy each of the three conditions for non-ideal justice (in the clinical sense) in isolation. However, to be non-ideally just, some particular use of SRM would need to satisfy all three of these conditions. Morrow and I expressed skepticism that some SRM policy would simultaneously satisfy the political feasibility and moral permissibility conditions, because these conditions seem to pull SRM in opposing directions.[29] One major reason for deeming SRM politically feasible is that it could be deployed by a single state or small coalition, sidestepping the collective action problem that makes it so difficult to achieve effective cuts in emissions. But unilateral or "minilateral" SRM is morally problematic. First, it would violate requirements of procedural justice, or the type of justice pertaining to how decisions ought to be made.[30] By definition, unilateral or minilateral SRM would be deployed without any decision-making input from those who are not among the deploying parties, despite the fact that SRM is likely to have global impacts, including harmful ones. Intuitively, this would be procedurally unjust to parties who are excluded from SRM decision-making. The next chapter is devoted to procedural justice, but for now we can note a plausible necessary condition for such justice: if some party stands to be substantially impacted by some global climate policy, then that party must be afforded an opportunity to contribute to decision-making about whether to enact that policy. Second, unilateralism or minilateralism could favor SRM policies that are distributively unjust, entailing disproportionate risks of harm to non-deploying parties. This is because members of a deploying coalition are likely to prefer policies that are beneficial for themselves, even if that means shifting substantial risks to others. If the interests of other parties, such as those of Indigenous peoples, are not represented in decision-making, then there is little reason to think that those interests will be afforded much concern.[31]

These considerations raise a serious concern. Although unilateral or minilateral SRM might be politically feasible, at least for sufficiently powerful parties, it is questionable whether such a use of SRM would be morally permissible, even in the relaxed sense of non-ideal theory. Given its potentially substantial injustice (both procedural and distributive), unilateral or minilateral SRM could carry moral ills that are not proportionate to the goods achieved. As noted above, there may be uses of SRM that, under certain conditions, would be morally permissible, and these are perhaps unlikely to involve unilateral or minilateral decision-making. However, such uses of SRM are not likely to be politically feasible. To satisfy our (intuitively plausible) necessary condition for procedural justice, SRM decision-making would need to include broad participation, given the global impacts and high stakes of

SRM. Yet this invites something akin to the collective action problem familiar from climate negotiations, as it could be very difficult to reach a decision to deploy SRM, given the potentially divergent interests of decision-makers. Different parties might have conflicting preferences regarding the intensity of cooling to be administered, the location of deployment, the timescale for deployment, other measures to accompany SRM (e.g., regarding adaptation or mitigation), and whether SRM should be used at all. In that case, SRM deployment could face impediments similar to those faced by attempts to achieve deep cuts in global emissions. Indeed, this potential for divergent interests provides a powerful, strategic incentive for SRM coalitions to remain small.[32] Because of this, there is a deep tension between the feasibility and permissibility conditions for non-ideal justice in the case of SRM, making it difficult for an SRM to simultaneously satisfy both.

We also identified a second reason for thinking there is tension between the permissibility and feasibility of SRM. Any reasonable proposal for SRM will allow that it should not constitute a permanent intervention in the climate system, for this would put future generations at risk of unjust harm indefinitely. It could be difficult for perpetual SRM to deliver goods that are proportionate to the ills it would carry into the indefinite future, especially once net emissions and atmospheric concentrations decline to safe levels. This raises the question of whether an SRM policy without an "exit strategy" would be morally permissible.[33] A viable exit strategy would specify how SRM is to be phased out as net emissions and atmospheric concentrations decline. Without such a strategy, present-day agents of SRM might commit future generations to maintaining SRM into the further future. Yet whatever the intentions of its framers, it is unclear that an SRM policy with a viable exit strategy would be politically feasible. Given the rapid warming (and attendant risks) that would occur after a sudden termination of SRM, a viable exit strategy must specify benchmarks for reducing the radiative forcing of SRM, as well as the rate at which SRM is to be gradually drawn down. It could be difficult to reach broad, multilateral agreement on such matters. This may be less difficult in the case of unilateralism or minilateralism, but as we have seen, such exclusionary decision-making raises concerns about moral permissibility.

Nonetheless, even supposing that it is politically feasible for decision-makers to reach agreement on how to phase out SRM, it is unclear that it would be feasible to reach the relevant benchmarks for initiating a cessation of SRM. The concern here is that emissions and atmospheric concentrations will continue to rise in the wake of SRM deployment. If SRM is effective in reducing various risks driven by rising emissions, then phasing it out would carry its own risks of harm and injustice. As Morrow and I noted, some commentators "fear that society will be unable to reduce emissions, especially once SRM eases the (prospective) pain of climate change. If that is true, to deploy SRM is effectively to commit the world to SRM in perpetuity."[34] Whether this is so depends partly on the controversial issue of whether SRM deployment poses some type of "moral hazard," such as by masking the effects of climate change and thereby reducing incentives to mitigate emissions.[35] There have been some empirical studies of the attitudes of individual members of the public toward emissions mitigation under SRM,[36] but there remains a great deal of uncertainty about how successful SRM might impact decision-making at a higher level. The fact that individuals say and (presumably) believe that SRM would not erode their commitment to mitigation is obviously no guarantee that the relevant institutions and states would be so committed—especially in the wake of successful deployment—even assuming they start off with sincere intentions to support mitigation prior to SRM deployment. To be clear, I am not assuming that SRM would in fact pose a moral hazard at an institutional or state level. We are not currently in a position to know that. But if SRM does pose such a moral hazard, that provides yet another reason to worry that political feasibility and moral permissibility pull in opposite directions.

While there is tension between SRM's ability to satisfy (simultaneously) the conditions of feasibility and permissibility, this does not entail that all SRM policies *must* be unjust in the clinical-theoretic sense. The moral permissibility condition is sensitive to the social and geophysical facts that happen to hold at any given time. Recall that the comparative criterion of that condition directs us to consider the relative merits and deficiencies of various politically feasible options. A given SRM policy might compare favorably to other feasible policies in the future even if it compares disfavorably at present. This is because some feasible, non-SRM policy (e.g., one involving emissions mitigation) might be effective in averting injustice at one time but ineffective in doing so at another time (e.g., once past emissions have committed us to substantially harmful impacts). Likewise, the moral ills of some politically feasible SRM policy might be proportionate to the goods it secures at a later time, even if that same policy fails to satisfy the proportionality criterion at the present time. Again, this is due to changing social and geophysical conditions. In chapters five and six, I will sketch what I call "pessimistic climate scenarios," or cases in which all available courses of action carry substantial moral problems, such as risks of severe harm and injustice. In such scenarios, I will argue, some politically feasible uses of SRM are plausibly taken to be morally permissible. In cases like that, the moral ills of anthropogenic climate change would be very great. Suppose that some SRM policies would be effective in averting many of those ills, but that the politically feasible variants of these policies all involve unilateral or minilateral decision-making, thus carrying moral ills of their own. Despite this, deployment of SRM might be permissible here, provided both that the goods secured (e.g., in the form of averting severe risks) are proportionate to the ills SRM carries *and* that this SRM policy provides the best ratio of goods secured to ills imposed when compared to other options that are feasible in the relevant scenario.

Admittedly, this is a very high bar to meet, but that fits with the conventional wisdom that SRM would be justified (if ever) only in extreme circumstances, as well as the common view among researchers that deployment at the present time would not be appropriate. Yet if we continue to make little progress on the mitigation front, it becomes more likely that humanity will face a pessimistic scenario in the future, and it is in such a scenario that some uses of SRM could be non-ideally just in the clinical-theoretic sense. At any rate, this is the argument I will make in chapter six.

Compensation for Unjust Harms

One way to mitigate some of the injustice of any climate policy is to provide economic compensation to those who have been unjustly harmed. Such compensation could go some distance in relieving injustices resulting from various policy measures.[37] Suppose, for example, that deployment of SRM reduces precipitation in some region, harming some innocent parties by making it more difficult for them to meet certain basic needs, such as for food or freshwater. In line with our convergence hypothesis, any plausible theory of distributive justice will identify this as an unjust harm, although they will provide competing accounts of why this is so. In some cases, compensation could ameliorate this unjust harm, providing financial means for some parties to meet their basic needs via other sources. All else being equal, deploying SRM with compensation for victims is axiologically preferable (from the point of view of justice) to SRM without such compensation. Nonetheless, compensation is not a cure-all for the potential injustices of climate policies. One reason for this is that not all unjust harms are susceptible to being redressed economically. Death is an obvious example of this. Other examples include the loss of irreplaceable goods, such as those dependent on culture, specific geographical locations (e.g., low-lying islands), and so on.

Even in the case of unjust harms that are in principle susceptible to economic remuneration, there remain substantial challenges to instituting an effective SRM compensation regime that is itself just. First, there are technical challenges arising from scientific uncertainty, particularly when it comes to detecting and attributing harmful impacts to SRM.[38] Such detection and attribution techniques are likely to be needed for an SRM compensation regime, partly because we cannot anticipate all the (potentially unjust) consequences of SRM in advance of its deployment.[39] Instead, we would need to identify these unjust harms as they arise. Fine-grained detection of particular impacts is likely to be difficult, given our limited historical data for various climate variables in some regions, limitations on current observations of such variables, and diversity among the techniques used to detect changes.[40] These limitations might prevent us from identifying particular impacts that have in fact occurred.

Perhaps more problematically, even if some impact is properly detected, in some cases it could be very difficult to attribute that impact to SRM. Suppose that SRM is deployed and that, several years later, a severe drought occurs (and is properly detected) in the Sahel region.[41] We would risk committing a *post hoc* fallacy if, on the basis of no further information, we inferred that SRM was the cause of this drought. While it is true that some uses of SRM carry risks of affecting precipitation patterns in ways that could result in drought, it is also true that droughts would sometimes occur in the absence of SRM. In this case, the question would be whether the drought under consideration was caused by SRM rather than "natural" or other anthropogenic factors. This can be challenging, because the answer to this question depends on a counterfactual, namely the conditions that would have held in the relevant region had SRM not been deployed. Because of this, we could not determine through observation alone whether SRM caused the drought. Here we must rely on (imperfect) climate models, simulating the conditions that would have held in this region absent SRM. If we find that the drought in question does not occur in simulations that lack SRM, then we have some basis for attributing the drought to SRM. Yet because such modeling is an uncertain—and, in the case of divergent modeling results, deeply uncertain—enterprise, it could often be unclear whether or not SRM is causally responsible for some set of harmful impacts.[42] In cases of substantial uncertainty regarding attribution, we might remain ignorant as to whether some parties warrant compensation, even if we have made sincere efforts to dispel such ignorance.

We might deal with this problem by appealing to the notion of "fractions of attributable risk."[43] Rather than identifying some particular cause of a given event, this technique determines the extent to which some factor (e.g., SRM deployment) makes a given event (e.g., drought in some region) more probable. While it could be very difficult for climate models to attribute some *specific* event to SRM, they may be able to show that events of the relevant kind are more likely to occur by some specified degree. If, for example, models show that some deployment of SRM makes the occurrence of drought in some region twice as likely to occur, then SRM's fraction of attributable risk for such an event in the future would be one-half. With this information, one could argue that victims of such droughts warrant SRM compensation, despite uncertainty as to whether some particular drought has been caused by SRM. One approach is to hold that such victims should receive full compensation for the harm they experience, provided that SRM's fraction of attributable risk for that harm exceeds some threshold (say, one-half).[44] Another option is to hold that victims should receive only partial compensation equivalent to the relevant fraction of attributable risk. On this approach, payees in the case above would be compensated for half the harm they experience from the drought in question. However, even putting aside the issue of whether partial or full compensation would be appropriate here, this technique is still plagued by uncertainty, as Irvine and I have noted elsewhere.[45] In order to calculate the appropriate fraction of risk attributable to SRM for some event, we would need to rely on extensive simulations of the climate, and this requires using imperfect computer models. This would be unavoidable, because we must compare the actual climate to a counterfactual (and therefore unobserved) climate, the latter of which can only be simulated. This opens the door for deep uncertainty regarding what fraction of attributable risk is appropriate to assign in some case, as different modeling studies might provide divergent probability estimates for a specific event. Accordingly, although approaches based on fractions of attributable risk are worth exploring further,[46] it is unclear how far they can go in reducing uncertainty about causal attributions.

The second set of challenges for crafting a just SRM compensation system owes to the substantial ethical uncertainty surrounding what would count as compensatory justice in such cases. We may treat compensatory justice as a subset of distributive justice, for the aim of the former is to rectify unjust distributions resulting from deployment of SRM. Even in cases in which harms are properly detected and correctly attributed to SRM, we still face three ethical questions: what parties ought to pay compensation, what parties deserve compensation, and how much compensation ought to be paid to deserving parties? There are competing, reasonable answers to each of these questions.[47] We might think of this as a case of Gardiner's "theoretical storm,"[48] as these issues are subject to ongoing debate within ethical theory. I address each of these questions below, arguing for an SRM compensation system in which payment is provided by those agents of SRM who are also high emitters of greenhouse gases, payment is received by anyone falling below an impersonal threshold of well-being as a result of SRM, and the amount of payment comes to whatever is needed to return such parties to that impersonal threshold. I will argue that, as a matter of non-ideal distributive justice, such a system likely offers our best option. Here we will observe an important virtue of taking a non-ideal-theoretic approach. Because it is limited to those options that are politically feasible and likely to be effective, many conceivable compensation systems are taken off the table, despite the possibility that some of them might be attractive for various moral reasons.[49] Accordingly a non-ideal approach need not resolve (possibly intractable) debates in ethical theory before making claims about the (non-ideal) justice or injustice of competing compensation schemes. This is tremendously useful, for it allows us to navigate Gardiner's theoretical storm much more effectively than those taking ideal approaches to climate justice.

The first question plausibly hinges on a familiar debate within the climate ethics literature regarding what parties ought to bear the costs (however conceived) of responding to climate change, whatever response is to be favored. Most positions draw on one of the following principles: Polluter Pays, Beneficiary Pays, or Ability to Pay. According to Polluter Pays, the agents of pollution (e.g., greenhouse gas emissions, sulfate aerosols) ought to bear the relevant costs. In the case of climate change, this might mean bearing economic burdens associated with cutting one's own emissions or paying for adaptation on the part of low emitters. If used as part of an SRM compensation system, responsibility for compensating appropriate parties unjustly harmed by SRM would fall to the agents of SRM, for such agents would be causally responsible for the pollution (e.g., sulfate aerosols) to which the unjust harm is attributable. Caney usefully distinguishes micro and macro versions of this principle. On the micro-version, "if an individual actor, X, performs an action that causes pollution, then that actor should pay for the ill effects of that action." On the macro-version, "if actors X, Y, and Z perform actions that together cause pollution, then they should pay for the cost of the ensuing pollution in proportion to the amount of pollution that they have caused."[50] Which version is relevant in the case of SRM would depend on what agents deploy it. In the case of unilateral deployment, the micro-version arguably suffices, given that there would be a single agent (e.g., some state) causally responsible for the unjust harms in question. In the case of multilateral deployment, the macro-version would be needed, dividing moral responsibility for compensation in accordance with the causal contributions made by the several agents of SRM, insofar as this can be determined.

Polluter Pays might seem very plausible here, but in addition to standard worries about it,[51] this principle is arguably inappropriate for certain uses of SRM. This is because SRM might be deployed in order to combat the substantial moral ills of anthropogenic climate change. Let us set aside considerations of political feasibility for the moment and imagine that SRM is deployed by a group of parties who are primarily concerned with reducing the injustice and overall harm of climate change. Suppose they are largely successful in this, slowing the rate of temperature change, buying time for poverty reduction via fossil fuels, and so on. Despite these generous suppositions, it is virtually certain that some parties would be harmed—in ways that are intuitively unjust—as a result of SRM, whatever baseline we use to determine harm (see below). On a straightforward application of Polluter Pays, the agents of SRM are required to compensate those harmed by SRM. These agents might object that this requirement is itself unjust. After all, as we are supposing for the moment, the overall injustice and harm would have been much worse absent SRM, so it might be unfair to require these agents of SRM to pay compensation. In fact, still in the spirit of Polluter Pays, they might argue that the agents of *greenhouse gas pollution* should be made to compensate those harmed by SRM, pointing out that it was only the intractability of such polluters that moved the agents of SRM to deploy it. Had greenhouse polluters cut their emissions as they ought to have done, then there would be no need for SRM, and hence no SRM-induced harm.

Of course, it is not likely that actual agents of SRM would have such moral concerns as their *primary* motivations. Perhaps more realistically, consider a low-emitting, small-island state with substantial poverty that joins some coalition in deploying SRM.[52] Suppose, as is plausible, that this state is motivated by the reasonable desire to preserve its own existence, hoping that SRM will help avert otherwise catastrophic (for the state in question) sea-level rise. According to Polluter Pays, this state would be morally responsible for compensating those harmed by SRM, and (on the macro-version) in proportion to the amount of SRM pollution that state has contributed. Once again, the small-island state might object, pointing out that it had little choice in the matter, as its existence was threatened by a problem caused by high emitters elsewhere. Arguably, it would be perverse to require this state to compensate those harmed by SRM, especially given its poverty and low emissions. Like the case of morally motivated agents of SRM, this case indicates that we should be cautious in using Polluter Pays for SRM compensation.

Nonetheless, requiring polluters to pay is plausibly taken to be just in some cases of SRM, particularly those involving what Gardiner calls "moral schizophrenia," or a state in which we endorse some ethical position that, in other contexts, we are unwilling to endorse.[53] As an example, imagine some high-emitting state that consistently undermines the global community's attempts to coordinate substantial cuts in emissions. Suppose this state then deploys SRM, arguing that it is needed to address the serious problems of climate change. On the one hand, this state has (at least implicitly) consistently denied the dangers of climate change, yet on the other it is now appealing to those very dangers as a justification for deployment of SRM. Unlike the small-island state just discussed, this state cannot argue that SRM is being deployed in response to a problem created by others, and so it seems just to require this state to pay compensation to those harmed by SRM. After all, it is only because of this state and others like it that SRM was (we are assuming for now) needed. Both these cases suggest that considering SRM compensation in isolation from the ills of climate change would be too restrictive. Whether agents of SRM ought to compensate those harms arguably depends in part on those agents' contribution to the problem of climate change. However, as Garcia has argued, some version of the Polluter Pays principle can be preserved despite these concerns.[54] Below, I will defend a version of Polluter Pays that understands polluters in a two-pronged sense, namely those agents of SRM who are also high emitters of greenhouse gases.

A second standard principle used in climate ethics is that of Beneficiary Pays.[55] Regarding SRM compensation, this principle would require those who benefit from SRM deployment to compensate those who are unjustly harmed by it. Unlike Polluter Pays, this principle takes causal responsibility to be irrelevant in determining moral responsibility for providing compensation. While the agents of SRM might benefit from it, the beneficiaries could also include many non-agents, including parties who opposed deployment, and Beneficiary Pays would require such parties to compensate victims of SRM injustice. At the same time, this principle could relieve some SRM agents of any obligation to compensate victims of unjust harm, for some of them might turn out not to be beneficiaries of deployment, presumably counter to their hopes.

At least for the purposes of SRM compensation, Beneficiary Pays runs into problems. Intuitively, it appears unfair to require non-agents of SRM to compensate victims of unjust harm, even if those non-agents have benefited from SRM. This would hold especially in cases in which beneficiaries of SRM had strongly opposed its deployment. Why should those who opposed—or, more weakly, those who did not consent—to deployment be held morally responsible for compensating victims of other parties' choices? Likewise, it appears unjust to let agents of SRM off the hook simply because they do not benefit from deployment. Imagine that some party deploys SRM with the intention of benefiting itself, disregarding grave risks to other parties, but that the actual outcomes result in slight harm to the deploying party while substantially harming others. Because, contrary to its goal, the deploying party is not an actual beneficiary of SRM, Beneficiary Pays does not require it to provide compensation to those who suffer substantial harms. This is problematic, because it allows reckless agents (e.g., those who ignore grave risks to others) to avoid compensating victims of their recklessness, provided that some factor (e.g., bad luck or irrational decision-making) prevents reckless agents from benefiting from their policies.

As I discuss below, any SRM compensation system will be subject to uncertainty when it comes to identifying those who have been harmed by deployment. Unlike the other principles, however, Beneficiary Pays is subject to further uncertainty, namely in identifying those who have benefited from deployment. While it is fairly straightforward to *identify* polluters and those able to pay, whether or not some party is a beneficiary of SRM could be obscured by scientific uncertainty regarding detection and attribution. It is not enough to observe that, in the wake of SRM, some party is better off relative to some baseline, for this might not be a result of SRM. Even if we know that someone is better off due to some climate-related phenomenon, this could be due to natural variability or other anthropogenic factors. As noted above, even with careful observation and reliance on improving climate models, there could be substantial uncertainty when it comes to attributing beneficial impacts to SRM. Putting aside ethical uncertainty regarding whether Beneficiary Pays is the right principle for SRM compensation, the scientific uncertainty just noted might make it very difficult to apply Beneficiary Pays in the real world. While we might be convinced that beneficiaries ought to pay, we simply might not know who the beneficiaries are.

Analogously, the Beneficiary Pays principle is subject to a more acute form of the non-identity problem than either of the other two principles. Suppose that SRM is deployed and maintained into the future. Because of its global impacts, affecting conditions and patterns of reproduction, SRM could result in the birth of many persons who are non-identical to those who would have been born had SRM not been deployed. This is potentially a problem for any compensation system, because such non-identity is arguably relevant to whether or not some future person has been harmed by SRM. It is at least a puzzle how someone whose existence depends on SRM could also be harmed by SRM. After all, had SRM not been deployed, it is not the case that such a person would be better off, for she would not exist absent SRM. So all SRM compensation systems face a potential non-identity problem for future recipients of compensation, a matter to which I return below. However, Beneficiary Pays also faces a non-identity problem on the side of those who are to provide compensation in the future. If some future person has a high standard of living due to the impacts of SRM, but his existence depends upon this past deployment of SRM, it is arguably the case that he is not a proper beneficiary of SRM. Once again, had SRM not been deployed, he never would have been born, so it is not the case that he is better off in the actual world of SRM than he would have been in some counterfactual world, provided that being brought into existence is not itself a benefit. This raises a concern that SRM compensation systems dependent upon Beneficiary Pays may have very limited application in the future, for it is unclear whether there would be proper beneficiaries (hence, compensators according to Beneficiary Pays) beyond the generation that deploys SRM. This would be a curious result. Although SRM could have harmful impacts for future generations, this implies that there may be no one to compensate them.

There are a variety of responses to the non-identity problem, of course. As already noted, we might adopt an impersonal baseline for determining whether some party has been harmed or benefited. In the present case, we might argue that some person has been benefited by SRM if and only if that person's well-being lies at or above some specified threshold because of SRM. In this case, we do not compare a person's well-being in the actual world to what it would have been absent SRM. Rather, we only consider whether someone's actual well-being meets the threshold in question and whether this has been caused (or perhaps allowed) by SRM. Here it does not matter whether someone's existence depends on the past deployment of SRM, for we do not take into account how well-off that person would have been in some counterfactual scenario. While this approach can avoid the non-identity problem, it faces two challenges: arguing for an appropriate impersonal baseline for benefits, and attributing such benefits to SRM.

Another potential solution to the non-identity problem is to take collectives, rather than individual persons, to be the proper subjects of harm and benefit.[56] Although many individuals born after SRM deployment are likely to be non-identical to those who would have been born otherwise, it is plausible to expect that some collective entities (e.g., communities, states, corporations) would not owe their existence to SRM. Thanks to SRM, some of these collectives might be better off than they would have been in some (but not necessarily all) counterfactual scenarios, and so we may treat them as proper beneficiaries of SRM deployment relative to those scenarios. On Beneficiary Pays, such collective entities would then owe compensation to victims of SRM-induced injustice. Importantly, identifying collectives as appropriate compensators does not require one to identify collectives as the (only) appropriate recipients of compensation. There is conceptual space to hold that collective entities are morally responsible for providing compensation to victims of injustice, but that it is individual persons (rather than collectives) who warrant that compensation.

Finally, Ability to Pay would require those with the capacity to compensate victims of injustice to provide such compensation. A standard version of this principle is that parties ought to provide compensation in direct proportion to the resources available to them.[57] In the case of SRM, Ability to Pay would require wealthy parties to compensate those unjustly harmed by deployment, regardless of whether those parties are either agents of SRM or beneficiaries of it. On the standard version of the principle just mentioned, such parties would compensate in accordance with their relative wealth of resources: parties with greater Ability to Pay would provide more SRM compensation than parties with less Ability to Pay.

Like the other principles, it is not difficult to envision applications of Ability to Pay that would be intuitively unjust to compensators. Imagine that SRM is deployed over the objections of some wealthy party, which suffers harm in the process but nonetheless retains ample resources. Ability to Pay would require this party to provide compensation to victims of SRM-induced injustice, but this party might reasonably object to this requirement, for it is neither morally nor causally responsible for the unjust impacts of SRM. In addition, this party opposed deployment of SRM and was harmed by that deployment. Although this is a hypothetical case, it is not unrealistic to suppose that *some* wealthy countries and their citizens might object to SRM and yet suffer some harm from its deployment. Requiring them to compensate victims of SRM would seem to be unfair.

Instead of treating the relevant polluters as equivalent to the agents of SRM, we might instead identify the relevant polluters as those agents of SRM who are also high emitters of greenhouse gases. This would situate SRM compensation within the broader context of climate change, allowing us to take into account a party's contribution to the problem that SRM would seek to address. For instance, relying on this conception of who counts as a polluter, the Polluter Pays Principle presumably would not require compensation to be provided by a small-island state that joins an SRM coalition, given the likelihood of low emissions in that state. However, this understanding of Polluter Pays would require compensation provision from members of an SRM coalition who have contributed substantially to the emissions driving climate change. So this approach can avoid the concern that Polluter Pays unfairly burdens low emitters. More broadly, this approach has the virtue of not treating SRM in isolation. This is appropriate, as no serious proponent of SRM research thinks deployment would be a good idea in its own right—rather, such proponents see deployment as a potentially fruitful means for managing the various ills of climate change. Because of this, it would be a mistake not to link SRM compensation to broader considerations about climate change and the emissions driving it.

We should consider a likely objection to this version of Polluter Pays. Suppose that SRM is deployed by a coalition comprised exclusively of low-emitting states. On the version of Polluter Pays just presented, none of these parties would be a polluter in the relevant sense, and so none of them would be obligated to provide compensation to victims of SRM-induced harm. This threatens to leave such victims in the lurch. If being a polluter in the two-pronged sense is a necessary condition for being obligated to provide SRM compensation, and if the agents of SRM are all low emitters, then no one is obligated to provide SRM compensation. However, as a geopolitical matter, it is very unlikely that the agents of SRM would be comprised entirely of low emitters. It is difficult to imagine that powerful states would decline to exert their influence over substantially less powerful states seeking to engineer the climate, particularly if there is a concern that this would run contrary to the perceived self-interests of some powerful states. This is simply in line with the truism that powerful parties are likely to exert influence over matters in which they have substantial stakes. As a general rule, low-emitting states tend to have less power on the global stage than high-emitting states. Although we can imagine an SRM coalition comprised solely of low-emitting states, that is not politically feasible in reality, for high-emitting states will have preferences regarding the deployment and maintenance of SRM. Some high emitters might seek to thwart such a coalition (e.g., through economic or military mechanisms). Given the power differential, high emitters are likely to succeed in this venture. Alternatively, some high emitters might seek to join and influence the coalition in question. In the first case, SRM is prevented, and so there is no need for SRM compensation. In the second case, SRM is deployed, but the agents behind it include high emitters. Such parties would be polluters in the two-pronged sense, and thus obligated to provide compensation on the corresponding version of Polluter Pays.

I have been referring vaguely to victims of SRM as being the recipients of compensation. We need a more careful consideration of the appropriate recipients of compensation. It might seem obvious that any party harmed by the impacts of SRM ought to be compensated, but this is not necessarily the case. Because we are investigating compensation as a way to ameliorate unjust distributions resulting from SRM, our focus should be on *unjust* harm, and not all instances of SRM-induced harm are plausibly taken to merit compensation. Consider a high emitter who is made slightly worse off economically by SRM, or consider an agent of SRM that harms itself due to a reckless deployment of SRM. Although both parties are harmed, arguably neither is a case of unjust harm, nor does either merit compensation.[58] In certain cases, it may be clear that some party has been harmed by SRM in a way that, in line with the convergence hypothesis, counts as unjust on virtually any plausible theory of justice. It is less obvious whether compensation would be deserved by those who miss out on benefits they would have enjoyed had SRM not been deployed. Although climate change is expected to be quite harmful on the whole, there may be individual beneficiaries of it (e.g., due to longer growing seasons in some regions). If SRM deployment cancels these benefits, those affected might request compensation, given that they are less well-off than they would have been. There is ethical uncertainty regarding whether such parties ought to be compensated. This issue is also connected to the technical challenges of detection and attribution discussed above. Whether some party is less well-off under SRM than it would have been without SRM is not easily determined, as it depends on the level of well-being this party would have enjoyed in a counterfactual case. We might attempt to determine this counterfactual via (imperfect) climate models, but this is subject to the problems noted above. So even if we decide that, in principle, compensation is deserved by parties who miss out on the benefits of climate change because of SRM, it could be very difficult to determine who those parties actually are. Now one might object that the same epistemic problems arise for detecting—and thus compensating—other types of SRM-induced harm, thus raising (potentially severe) challenges even for SRM compensation systems that do not seek to remunerate missed benefits. Partly in response to this potential problem, I will defend the idea of using an impersonal baseline for compensation, providing payment to those who (unjustly) fall below some threshold of well-being as a result of SRM. As we shall see, this allows us to avoid many of the technical problems that plague counterfactual approaches. By taking such an impersonal approach, the goal of an SRM compensation system would be to ensure that parties, regardless of their identity, do not unjustly fall below some threshold of well-being. As we shall see, because this approach does not try to compensate parties with respect to their historical or counterfactual levels of well-being, the non-identity problem can be sidestepped.

Finally, there are questions about how much compensation should be provided to the appropriate recipients. This depends on the baseline that is used. In general, there are two kinds of baseline we might employ. First, we might use some historical baseline, providing compensation payments to a victim of SRM, in order (presumably) to return her to a level of well-being she enjoyed at some point in the past. On this approach, after identifying some victim of SRM, we would determine the level of well-being this party enjoyed prior to SRM's deployment (or prior to some impact thereof), comparing it to the lower level of well-being she now has. Compensation would then be provided to return this party's well-being to the level of the historical baseline. In other types of case, this may be a reasonable approach to compensation. For instance, if someone's property is destroyed through the malicious action of another, then the victim presumably warrants compensation in an amount equal to the value of the destroyed property, effectively returning the payee to her previous level of well-being. But this is not a helpful approach in the case of SRM. Once again, the problem is that a purely historical baseline threatens to decouple SRM from the broader context of climate change. Arguably, historical baselines are less relevant here, because climate change will alter the levels of well-being of many persons—in many cases, by substantially reducing those levels. SRM could be used in an attempt to limit such reductions, and it *might* be successful in so doing. Still, no one believes that SRM would be a cure-all in this venture. Even on the most optimistic prognoses, SRM deployment is likely to have harmful side effects.

Consider a case in which SRM is deployed, resulting in regional changes in precipitation that are clearly attributed to SRM. Imagine a farmer who is affected by these changes, such that the agricultural productivity of her land is reduced relative to what it was in past years, such that she is economically less well-off than in the past. The backward-looking historical approach to compensation would suggest that this person deserves SRM compensation, because the impacts of SRM have made her worse off than she was historically. But it might be that, had SRM not been deployed, she would have been worse-off still as a result of climate change (e.g., due to heat stress on her crops). Using a purely historical baseline for compensation would not take into account the future harms of climate change that SRM allows us to avoid. This seems like a mistake. In many cases, it might be practically impossible to preserve historical levels of well-being for some persons, given what climate change portends. In such a case, if some intervention limits, but does not fully avoid, a reduction in someone's well-being (relative to a historical baseline), it seems inappropriate to hold that the party in question warrants compensation for that intervention, for this ignores the broader context in which the intervention was undertaken.

This problem is avoided by the second approach, which is to use a counterfactual, forward-looking baseline. Rather than compensating victims of SRM in accordance with some actual level of well-being they have enjoyed in the past, this approach would compensate victims of SRM in accordance with what their level of well-being would have been had SRM not been deployed. In principle, this approach does not inappropriately award SRM compensation in cases like that of the farmer above, who is worse off than she was historically due to SRM but better off than she would have been without SRM. The counterfactual approach is inherently sensitive to the broader context of climate change, for it pays attention to how persons would have fared without SRM. This might seem like the right kind of approach, but using a counterfactual baseline is more prone to technical challenges than a historical approach, because it requires us to simulate scenarios that never occurred, whereas the historical approach can appeal to the conditions that actually held prior to deployment. There is a great deal of uncertainty when it comes to simulating counterfactual scenarios (e.g., by using an ensemble of modeling runs), and this epistemic difficulty could make it challenging to find a counterfactual baseline in which we can be reasonably confident.[59]

Moreover, there is ethical uncertainty regarding what counterfactual baseline we ought to use. There are many other trajectories that might occur should SRM not be deployed. Which one would occur is dependent on many factors that are difficult or impossible to predict, including the long-term impact of the Paris Agreement, the cost of renewable energy, whether or not some CDR technologies are utilized (and at what scale), what actual climate sensitivity turns out to be, and of course the total emissions produced by future humans. Of these many counterfactual possibilities, which one should we choose for the purpose of assessing harms? It is difficult to see how any such choice could avoid being arbitrary. This is unfortunate for the counterfactual approach, because this choice could make a substantial difference for assessing SRM's harms and benefits. Relative to a counterfactual scenario with a very high atmospheric concentration of greenhouse gases, a given party might be substantially better off with SRM, but that same party might be substantially worse off relative to a counterfactual scenario in which the atmospheric concentration is significantly lower than the first scenario. Moreover, regardless of what counterfactual baseline is used, the time at which we compare someone's well-being to what it would have been also could matter a great deal. Suppose that SRM is deployed to stave off climate-related harms that are clustered in the future. In that case, some party might be worse off (relative to the counterfactual baseline) two years after deployment, but better off (relative to the same baseline) twenty years after. Does this party warrant SRM compensation at year two?

Both kinds of baseline encounter questions when it comes to how much to compensate future persons. First, there is the question of discounting the future. If we discount future well-being as such, then arguably future persons will warrant less compensation than present persons, even if everything else (e.g., the impacts of SRM) is held equal. The exact reduction in compensation to future persons would be determined by the particular discount rate that is used. A similar point arguably applies, albeit less directly, for discounting future commodities (rather than well-being proper). One way that future persons can be harmed is through damage to the commodities they wish to consume. If SRM results in such damage in the future, but the value of the damaged commodities is discounted (perhaps steeply), then presumably future parties will receive less compensation than they would have received absent such discounting. Now one might object to what I have just said by noting that, by the time such damage occurs, those affected by it will no longer be future parties—rather, they will be present parties. Hence, neither their well-being proper nor the commodities they wish to consume should be subject to discounting. However, concerns about discounting are still relevant here, because some (perhaps much) of the funding for an SRM compensation regime will come from parties existing prior to some of the recipients of compensation. From the point of view of payers existing at the time SRM is deployed, some recipients of compensation will be parties who do not yet exist. If those payers discount the well-being of future persons or the value of the future commodities they will consume, this could affect how much funding is set aside for purposes of future payments. This is particularly salient for views that identify the appropriate payers as the initial agents of SRM, because these payers (whether individual persons or states) may cease to exist in the future, and so they might need to pay compensation in advance (e.g., into a long-term fund).

We should also note that, for persons born after the deployment of SRM, there are problems related to our choice of baseline for determining how much compensation is to be provided. There will be no historical baseline for well-being that applies, for the simple reason that there is no fact of the matter regarding such persons' levels of well-being prior to deployment, given that they did not exist at the relevant time. This aspect of the non-identity problem poses issues for both kinds of baseline. For future persons who owe their existence to SRM, similarly, there will be no counterfactual baseline that applied for such persons—again, because there is no fact of the matter regarding how well-off they would have been absent SRM, for they would not have existed in such cases. At best, this would be a severe limitation on SRM compensation systems employing such baselines, as this would seem to exclude many parties who (at least intuitively) warrant compensation payments.

We can avoid many of these problems by relying on an impersonal baseline, which would specify some threshold of well-being below which it is harmful for *anyone* to fall, regardless of one's historical identity or the conditions of one's existence. Those who are below this threshold due to SRM are taken to have been harmed by SRM. Someone falls below the threshold "due to SRM" if SRM played a causal role in bringing about the conditions or events that figure in the correct causal explanation of why that party is below the threshold. For instance, some party might live in poverty because of poor growing conditions brought about by a drought. Suppose that the correct explanation for why this drought occurred is that SRM had been deployed. Intuitively, the party in question has been harmed by SRM. This is so even if that party owes her existence to SRM's deployment. The crucial point is that, because of conditions caused by SRM, this party has a level of well-being below that specified by the threshold of the impersonal baseline.

Unlike historical and counterfactual baselines, this does not require us to compare some party's well-being post-deployment to what it was or would have been at a different time. Instead, we compare some party's actual well-being to the specified threshold. If this party's well-being falls below that threshold, and if this is attributable to SRM (in the sense specified in the previous paragraph), then we take this party to have been harmed by SRM. If this harm is of the unjust variety (e.g., because the party in question is a low emitter who is not an agent of SRM), then this party warrants compensation in the amount needed to return him to the level of well-being specified by the impersonal baseline. This avoids the problem faced by historical approaches. In a changing climate, levels of well-being are sure to change for many parties relative to their historical levels, and so the latter are arguably irrelevant here. The impersonal baseline is not backward-looking in this way, for it utilizes a minimal level of well-being to be maintained for all persons. For a different reason, this impersonal approach avoids many of the technical challenges of the counterfactual approach.[60] Although, ethically speaking, the counterfactual approach seems attractive, in many cases it could be virtually impossible to apply, given the vast uncertainty regarding how well-off some party would have been had SRM not been deployed in the past. Finally, as we noted above, an impersonal baseline for harm also sidesteps the non-identity problem, a major advantage when it comes to compensating future persons who owe their existence to SRM's past deployment. As we have seen, such a person would have no historical or counterfactual well-being to which we could appeal. This is not problematic for an impersonal approach. All else being equal, the fact that these problems are avoided gives us reason to prefer an impersonal approach to historical or counterfactual approaches.

Of course, the impersonal approach is not without some difficulties of its own. First, it will be controversial to specify an exact threshold that is appropriate. Plausibly, the impersonal baseline should include meeting basic human needs for food, water, shelter, social stability, and the like.[61] Whether the threshold should be set higher (e.g., to the point needed to secure basic capabilities) than merely meeting basic needs is an important question. I will return to this matter below when discussing compensation in terms of non-ideal justice, because more demanding thresholds might not be politically feasible. In that case, although ideal justice might require compensation that is well above what is needed to meet basic needs, non-ideal justice might require substantially less compensation.

Second, there is a concern that this approach might be unfair either to compensators or to recipients of compensation. This is partly related to how high or low the threshold is set. If we set it high, there is a risk that compensators will be made to pay more than is actually appropriate. If we set the threshold too low, there is a risk that recipients of compensation will receive less payment than is appropriate. Suppose that we get the threshold right, however, such that neither of these risks is realized. There is still the potential for unfairness to both groups, because (one might argue) the impersonal baseline ignores the broader context of climate change. On the one hand, some party harmed by SRM might have been even worse off under some (or even all) other scenarios. In that case, compensators might allege that it is unfair that they be required to compensate the harmed party. After all, relative to at least some non-SRM scenarios, this party would have benefited from SRM. All else being equal, the higher we set the threshold for the impersonal baseline, the more likely it is that compensators will be made to overpay. On the other hand, under some (or even all) other scenarios, some party harmed by SRM might have enjoyed a level of well-being that is far above the threshold specified in the impersonal baseline. In that case, the party in question might claim that the amount of compensation received is insufficient, because it does not fully remunerate this party's reduction in well-being (relative to what it would have been under some other scenario). All else being equal, the lower we set the threshold for the impersonal baseline, the more likely it is that recipients of compensation will be underpaid in this way.

There are several things to note about this second set of challenges for an impersonal baseline. First, because these concerns are predicated on counterfactuals (i.e., levels of well-being under scenarios that do not occur), they do not apply for future parties who would not have existed had SRM not been deployed. Depending on how long SRM is maintained, this could count for a large portion of those who are to be covered by a compensation regime. Accordingly, these concerns are limited to the portion of parties who would have had some level of well-being under alternative scenarios. However we set the impersonal threshold, there is indeed a risk of over- and undercompensating some parties. As we have seen, however, determining counterfactual levels of well-being is beset by very difficult epistemic problems. In many cases, it might be virtually impossible to determine whether some party would have existed under some set of alternative conditions. The same holds for determining what that person's level of well-being would have been under those alternative conditions. We have also seen that it is controversial what set of alternative conditions we should use for purposes of comparison, given the many possible scenarios that might have played out if SRM had not been deployed. Because of problems like this, an approach to compensation that requires us to take account of counterfactual well-being is likely to be impracticable. Here the impersonal approach has a major advantage, because it requires only that we compare some party's actual well-being to the specified threshold. In some cases, this approach might result in a level of well-being that is above or below what someone would have had in another scenario, but this is a tolerable risk, given that an impersonal approach avoids many of the severe practical problems of a counterfactual approach. We might think of the impersonal approach as providing a kind of "safety net," ensuring that victims of SRM injustice do not fall to intolerable levels of well-being.

Third, there are still technical challenges regarding detection and attribution when it comes to applying an impersonal baseline. For example, we might identify a case in which some party has fallen below the specified threshold due to a drought, but it might be uncertain whether this drought is attributable to SRM rather than natural variability. But this is a problem that will arise for any approach that seeks to compensate for SRM-induced harms. As we have seen, there are ways to deal with this problem, such as employing fractions of attributable risk to determine liability or utilizing improved climate models in the future to reduce uncertainty regarding attribution. Regardless, because this challenge will be faced by any SRM compensation system, it does not give us reason to favor some other approach over the impersonal one. Further, as we have observed, an impersonal approach sidesteps some of the technical challenges faced by other approaches, such as uncertainty regarding counterfactual levels of well-being. On the whole then, despite some remaining challenges, the impersonal approach fares *comparatively* well on the technical front. The only way to avoid many of these technical problems of attribution is to employ a general climate compensation system, in which payment is provided for climate damages regardless of cause.[62] On such an approach, we need not determine whether some climate-related harm is due to SRM, anthropogenic emissions, or natural variability, for the victim is equally deserving of compensation (or not, as the case may be) on all of these causes. There are two problems for such an approach. First, it could easily impose intuitively unfair burdens like those discussed above. Why should opponents of SRM be required to compensate for damages that (as it may be) are driven by SRM, and why should anyone be required to compensate victims of genuinely natural causes?[63] This potential for unfair burdens is problematic from both ideal and non-ideal perspectives on justice, the latter because unfair burdens would put pressure on the moral permissibility condition. Second, a general climate compensation system is unlikely to be politically feasible, as compensators may be unwilling to remunerate harms for which they are not causally responsible.

Because of these problems with a general approach, we are better off employing a compensation system that addresses SRM-induced harms. Any SRM compensation system is likely to be imperfect, but that does not mean that we should abandon compensation. If SRM is deployed, it is virtually certain that some parties will be harmed in ways that are (intuitively) unjust. All else being equal, imperfect compensation for unjust harms is better than no compensation. Accordingly, we should want an SRM compensation regime that, although imperfect, performs better than other options in the task of successfully remunerating unjust harms. I have argued that this bill is fit by a compensation system incorporating Polluter Pays (with polluters understood in the two-pronged sense), compensating parties that unjustly fall below some impersonal baseline of well-being (including, but not necessarily limited to, allowing parties to meet basic human needs) as a result of SRM, and compensating such parties in amounts needed to return them to that baseline.

It remains to consider whether such a compensation system would be favored by the non-ideal approach to justice that I am taking in this book. Recall that some policy is non-ideally just if it is politically feasible, likely to be effective, and morally permissible. Let us begin with the effectiveness condition. Because of its avoidance of many epistemic problems, the approach to compensation I favor here is more likely to be effective than the other approaches I have considered. In principle, it is a straightforward matter to distinguish polluters in the two-pronged sense from other parties, for we need only identify what parties are both agents of SRM and high emitters of greenhouse gases. As we have seen, identifying compensators could be very difficult on a Beneficiary Pays approach, because it may be unclear whether some party has benefited from *SRM* rather than natural variability in the climate, making it difficult to avoid mistakes when it comes to identifying actual beneficiaries. Accordingly, there is a concern that a Beneficiary Pays regime will be less effective than Polluter Pays in achieving just outcomes, because it might result in some non-beneficiaries being made to provide compensation while some actual beneficiaries evade that responsibility. On this point, Polluter Pays and Ability to Pay are roughly equal, as neither one faces an attribution problem in determining what actual parties qualify as compensators on their respective accounts. As I have noted, it is true that my approach to compensation faces an attribution problem in identifying what parties qualify as appropriate recipients of compensation, but this problem will be faced by any compensation system that seeks to remunerate victims of SRM. Finally, as we have just seen, utilizing an impersonal baseline to determine how much compensation ought to be paid avoids many of the epistemic problems faced by historical and counterfactual baselines. By avoiding the non-identity problem and uncertainty about counterfactual levels of well-being, an impersonal approach is more likely to be effective in administering compensation to victims of SRM.

There are also good reasons to think that a compensation system employing Polluter Pays and an impersonal baseline is politically feasible. Many existing compensation regimes already utilize Polluter Pays or something resembling it, rather than the more controversial Beneficiary Pays or Ability to Pay principles.[64] The last of these would be particularly contentious in the context of SRM. It is very unlikely that wealthy parties would be willing to pay (very much) to alleviate harms imposed by the actions of others. We might wonder whether my preferred type of compensation system is feasible when it comes to the amount of compensation that is to be paid. Presumably, this would be a major political sticking point, as even acknowledged polluters (in the two-pronged sense) will be unwilling to pay past a certain point. As a general rule, there is an inversely proportional relation between the political feasibility of some compensation system and the amount of compensation to be provided, all else being equal. Thus, the political feasibility condition can be expected to drive down the amount of compensation available to victims of SRM. This raises the concern that political constraints will make an ideally just compensation system infeasible, but this also has practical advantages when it comes to navigating ethical uncertainty. As noted above, it is uncertain where we ought to set the threshold of the impersonal baseline. As a matter of ideal justice—hence unconstrained by considerations of political feasibility—there are many options on which various parties might reasonably disagree. In the context of climate change, however, many of these options are likely to be politically infeasible, as compensators might be unwilling to provide payments in the amounts needed by some of these options. From the perspective of non-ideal justice, this takes those options off the table, so we may set them aside when it comes to crafting a non-ideally just compensation regime.

My suggestion is that the impersonal baseline should be set as high as is politically feasible. This is unlikely to be unfair to polluters, because they are likely to be unwilling to provide payments that are unduly burdensome to themselves. Within the constraints of non-ideal theory, this approach also yields the greatest amount of compensation one can hope to get from polluters, giving us the best chance to alleviate a great deal of any harm imposed by SRM. Whatever the maximal, politically feasible figure turns out to be, it is to be hoped that it would be sufficient at least to meet the basic needs of those harmed by SRM. There is no guarantee that this would happen, of course. It might be that the highest politically feasible baseline results in paltry compensation that is far too little to meet the basic needs of payees. This could happen for various reasons, such as if a small SRM coalition simply lacks the resources to provide much in the way of compensation. A case like this would give us good reason to disfavor deployment of SRM, but we are not yet in a position to examine that argument. At present, I am only considering what general *type* of compensation system would be favored under non-ideal theory.

This leaves us with the question of moral permissibility. Recall that some policy is non-ideally permissible if it satisfies both a proportionality criterion and a comparative criterion. The type of compensation system I have suggested does well on the proportionality criterion, because the good achieved (i.e., alleviating SRM-induced harms) seems proportionate to the burdens imposed (i.e., politically feasible payments on the part of high emitters who are also agents of SRM). As we saw above, other approaches risk imposing greater burdens in order to ameliorate harm. This is important for the comparative criterion. For instance, Beneficiary Pays and Ability to Pay risk imposing (intuitively) unfair burdens on parties not causally responsible for SRM, including parties who may have opposed SRM. A standard utilization of Polluter Pays risks imposing unfair burdens on agents of SRM who happen to be low emitters acting in self-defense against climate-related harms created by others. Now these intuitively unfair burdens could be justified on non-ideal approach, but not if there is some other option that carries a better ratio of goods achieved to ills imposed. The type of compensation system I have suggested performs better in this regard. The burdens it imposes on compensators are not (intuitively) unfair, and yet it still achieves an important good.

Notes

1. Jim M. Haywood et al., "Asymmetric Forcing from Stratospheric Aerosols Impacts Sahelian Rainfall," *Nature Climate Change* 3, no. 7 (2013): 660–65.
2. Andy Jones et al., "The Impact of Abrupt Suspension of Solar Radiation Management (Termination Effect) in Experiment G2 of the Geoengineering Model Intercomparison Project (GeoMIP)," *Journal of Geophysical Research: Atmospheres* 118, no. 17 (2013): 9743–52.
3. Marlos Goes, Nancy Tuana, and Klaus Keller, "The Economics (or Lack Thereof) of Aerosol Geoengineering," *Climatic Change* 109, nos. 3–4 (2011): 719–44.
4. Konrad Ott, "Might Solar Radiation Management Constitute a Dilemma?," in *Engineering the Climate: The Ethics of Solar Radiation Management*, ed. Christopher Preston (Lanham: Lexington Books, 2012), 40.
5. Toby Svoboda et al., "Sulfate Aerosol Geoengineering: The Question of Justice," *Public Affairs Quarterly* 25, no. 3 (2011): 157–80.
6. This name is adapted from Norton's convergence hypothesis, which predicts that anthropocentrists and non-anthropocentrists will tend to converge in their environmental policy preference. See Bryan G. Norton, *Toward Unity among Environmentalists* (Oxford: Oxford University Press, 1994).
7. For a helpful overview of these (and other) notions of distributive justice, see Julian Lamont and Christi Favor, "Distributive Justice," in *The Stanford Encyclopedia*

of Philosophy, ed. Edward N. Zalta (Stanford: Metaphysics Research Lab, 2016), https://plato.stanford.edu/archives/win2016/entries/justice-distributive/.

8. Richard J. Arneson, "Equality and Equal Opportunity for Welfare," *Philosophical Studies* 56, no. 1 (1989): 85.

9. Ibid., 86.

10. Ronald Dworkin, "What Is Equality? Part 1: Equality of Welfare," *Philosophy & Public Affairs* 10, no. 3 (1981): 185–246; Ronald Dworkin, "What Is Equality? Part 2: Equality of Resources," *Philosophy & Public Affairs* 10, no. 4 (1981): 283–345.

11. Amartya Sen, *Choice, Welfare and Measurement* (Cambridge: Cambridge University Press, 1982), 369.

12. John Rawls, *A Theory of Justice, Revised Edition* (Cambridge: Harvard University Press, 1999), 11.

13. Ibid., 53. Later in *A Theory of Justice,* Rawls provides a more detailed "final statement" of these principles of justice. See ibid., 266–67. However, the more commonly discussed statements I have provided here will be sufficient to guide our discussion of Rawlsian distributive justice.

14. For instance, I have not covered libertarian theories. See Robert Nozick, *Anarchy, State, and Utopia* (Basic Books, 1974).

15. To be clear, Rawls himself takes his two principles to hold for domestic justice (e.g., justice within some state) rather than for global justice. However, because climate justice is a global matter, it is reasonable to ask what would follow from applying Rawls' well-known principles at the global level.

16. Toby Svoboda, "Solar Radiation Management and Comparative Climate Justice," in *Climate Justice and Geoengineering: Ethics and Policy in the Atmospheric Anthropocene,* ed. Christopher Preston (Lanham: Rowman & Littlefield International, 2016): 3–14.

17. Eric Posner, "You Can Have Either Climate Justice or a Climate Treaty: Not Both," *Slate,* November 19, 2013, www.slate.com/articles/news_and_politics/view_from_chicago/2013/11/climate_justice_or_a_climate_treaty_you_can_t_have_both.html.

18. For a useful discussion of six ways of responding to such non-compliance, see Simon Caney, "Climate Change and Non-Ideal Theory: Six Ways of Responding to Non-Compliance," in *Climate Justice in a Non-Ideal World,* ed. Clare Heyward and Dominic Roser (Oxford: Oxford University Press, 2016), 21–42.

19. Rawls, *A Theory of Justice, Revised Edition.*

20. Darrel Moellendorf, "Taking UNFCCC Norms Seriously," in *Climate Justice in a Non-Ideal World,* ed. Clare Heyward and Dominic Roser (Oxford: Oxford University Press, 2016), 107, 110.

21. David Wiens, "Prescribing Institutions without Ideal Theory," *Journal of Political Philosophy* 20, no. 1 (2012): 45–70.

22. David R. Morrow and Toby Svoboda, "Geoengineering and Non-Ideal Theory," *Public Affairs Quarterly* 30, no. 1 (2016): 85–104.

23. Jason J. Blackstock et al., "Climate Engineering Responses to Climate Emergencies," http://arxiv.org/pdf/0907.5140; T.M.L. Wigley, "A Combined Mitigation/Geoengineering Approach to Climate Stabilization," *Science* 314, no. 5798 (2006): 452–54.

24. Scott Barrett, "The Incredible Economics of Geoengineering," *Environmental and Resource Economics* 39, no. 1 (2008): 45–54.

25. Blackstock et al., "Climate Engineering Responses to Climate Emergencies"; David Keith, *A Case for Climate Engineering* (Cambridge, MA: MIT Press, 2013).

26. David G. Victor, "On the Regulation of Geoengineering," *Oxford Review of Economic Policy* 24, no. 2 (2008): 322–36.

27. Morrow and Svoboda, "Geoengineering and Non-Ideal Theory," 93.

28. Ibid.

29. Scott Barrett et al., "Climate Engineering Reconsidered," *Nature Climate Change* 4, no. 7 (2014): 527–29.
30. Svoboda et al., "Sulfate Aerosol Geoengineering: The Question of Justice."
31. Kyle Powys Whyte, "Now This! Indigenous Sovereignty, Political Obliviousness and Governance Models for SRM Research," *Ethics, Policy & Environment* 15, no. 2 (2012): 172–87.
32. Katharine L. Ricke, Juan B. Moreno-Cruz, and Ken Caldeira, "Strategic Incentives for Climate Geoengineering Coalitions to Exclude Broad Participation," *Environmental Research Letters* 8, no. 1 (2013).
33. Christopher J. Preston, "Climate Engineering and the Cessation Requirement: The Ethics of a Life-Cycle," *Environmental Values* 25, no. 1 (2016): 91–107.
34. Morrow and Svoboda, "Geoengineering and Non-Ideal Theory," 95.
35. Benjamin Hale, "The World That Would Have Been: Moral Hazard Arguments against Geoengineering," in *Engineering the Climate: The Ethics of Solar Radiation Management*, ed. Christopher J. Preston (Lanham: Lexington Books, 2012), 113–31; Jesse Reynolds, "A Critical Examination of the Climate Engineering Moral Hazard and Risk Compensation Concern," *The Anthropocene Review* 2, no. 2 (2014): 174–191.
36. Adam Corner et al., "Messing with Nature? Exploring Public Perceptions of Geoengineering in the UK," *Global Environmental Change* 23, no. 5 (2013): 938–47.
37. Martin Bunzl, "Geoengineering Harms and Compensation," *Stanford Journal of Law, Science & Policy* 4 (2011): 70–76.
38. Toby Svoboda and Peter J. Irvine, "Ethical and Technical Challenges in Compensating for Harm Due to Solar Radiation Management Geoengineering," *Ethics, Policy & Environment* 17, no. 2 (2014): 157–74.
39. Clare Heyward, "Benefiting from Climate Geoengineering and Corresponding Remedial Duties: The Case of Unforeseeable Harms," *Journal of Applied Philosophy* 31, no. 4 (2014): 405–19.
40. D. A. Stone et al., "The Detection and Attribution of Human Influence on Climate," *Annual Review of Environment and Resources* 34 (2009): 1–16.
41. Haywood et al., "Asymmetric Forcing from Stratospheric Aerosols Impacts Sahelian Rainfall."
42. Svoboda and Irvine, "Ethical and Technical Challenges in Compensating for Harm Due to Solar Radiation Management Geoengineering."
43. Myles Allen, "Liability for Climate Change," *Nature* 421, no. 6926 (2003): 891–92; Joshua Horton, Andrew Parker, and David Keith, "Liability for Solar Geoengineering: Historical Precedents, Contemporary Innovations, and Governance Possibilities," *NYU Environmental Law Journal* 22 (2014): 225; P.A. Stott, D.A. Stone, and M.R. Allen, "Human Contribution to the European Heatwave of 2003," *Nature* 432, no. 7017 (2004): 610–14.
44. Horton, Parker, and Keith, "Liability for Solar Geoengineering: Historical Precedents, Contemporary Innovations, and Governance Possibilities."
45. Svoboda and Irvine, "Ethical and Technical Challenges in Compensating for Harm Due to Solar Radiation Management Geoengineering," 162.
46. Joshua Horton, "Solar Geoengineering: Reassessing Costs, Benefits, and Compensation," *Ethics, Policy & Environment* 17, no. 2 (2014): 175–77.
47. Svoboda and Irvine, "Ethical and Technical Challenges in Compensating for Harm Due to Solar Radiation Management Geoengineering."
48. Stephen M. Gardiner, "A Perfect Moral Storm: Climate Change, Intergenerational Ethics and the Problem of Moral Corruption," *Environmental Values* 15, no. 3 (2006): 397–413.
49. As an aside, I should note that compensatory justice falls within the realm of non-ideal theory, because in the case of full compliance with our ideal duties of justice there would be no unjust harm requiring compensation.

50. Simon Caney, "Cosmopolitan Justice, Responsibility, and Global Climate Change," *Leiden Journal of International Law* 18, no. 4 (2005): 753.
51. Ibid., 574.
52. Svoboda and Irvine, "Ethical and Technical Challenges in Compensating for Harm Due to Solar Radiation Management Geoengineering."
53. Stephen M. Gardiner, "Ethics, Geoengineering and Moral Schizophrenia: What's the Question?" in *Climate Change Geoengineering: Philosophical Perspectives, Legal Issues, and Governance Frameworks*, ed. William C. G. Burns and Andrew Strauss (Cambridge: Cambridge University Press, 2013), 11–38.
54. Robert K. Garcia, "Towards a Just Solar Radiation Management Compensation System: A Defense of the Polluter Pays Principle," *Ethics, Policy & Environment* 17, no. 2 (2014): 178–82.
55. Christian Baatz, "Responsibility for the Past? Some Thoughts on Compensating Those Vulnerable to Climate Change in Developing Countries," *Ethics, Policy & Environment* 16, no. 1 (2013): 94–110.
56. Edward Page, "Intergenerational Justice and Climate Change," *Political Studies* 47, no. 1 (1999): 53–66.
57. Henry Shue, "Global Environment and International Inequality," *International Affairs* 75, no. 3 (1999): 531–45.
58. Regarding the latter, we might have an obligation to provide aid, but this would be independent of any duty to provide compensation for unjust harm.
59. Svoboda and Irvine, "Ethical and Technical Challenges in Compensating for Harm Due to Solar Radiation Management Geoengineering."
60. Strictly speaking, my approach is counterfactual in one sense, but not in the sense of the counterfactual baselines discussed so far. Unlike those other baselines, on my view it does not matter what some party's well-being actually would have been had SRM not been deployed, and so my view is not counterfactual in *this* sense. Nonetheless, we might imagine what some party's well-being would have been absent some particular effect of SRM. For example, we might judge that persons in some region would be better off had SRM not led to drought in their region. Here we are imagining a counterfactual scenario—namely, the world as it actually is, except without drought in this particular region—to guide judgments of harm. Importantly, because it is impersonal, my approach does not require that this imagined scenario be one that could have held given some prior state of the world. Conversely, the counterfactual baselines I have been discussing compare someone's actual well-being to what that same person's well-being would have been under some other climate policy that could have been adopted. This is the crucial difference between my impersonal approach and those employing such counterfactual baselines.
61. For a helpful discussion of distinguishing needs from mere "wants" in mitigation policy, see David R. Morrow, "Wants and Needs in Mitigation Policy," *Climatic Change* 130, no. 3 (2015): 335–45.
62. Even a general climate compensation system is unlikely to avoid attribution problems in full. There could be cases in which it is uncertain whether some harm is climate-related.
63. Regarding the latter, there may well be duties on the part of some to provide aid to victims of natural variability in the climate, but these are not plausibly taken to be duties of *compensation*.
64. Eric Thomas Larson, "Why Environmental Liability Regimes in the United States, the European Community, and Japan Have Grown Synonymous with the Polluter Pays Principle," *Vanderbilt Journal of Transnational Law* 38 (2005): 541.

3 Decisions[1]

So far we have considered climate policy options in light of their potential outcomes. We have done this along two axes: overall harms and benefits, and the distribution (both spatially and temporally) of harms and benefits among various parties. Assuming we have both duties of beneficence and duties of justice, we ought to look for climate policies that promote or maintain both distributively just states of affairs and reasonable levels of overall well-being. In addition to these aims, we should also care about how decisions regarding global climate policy are made. By a *global* climate policy as opposed to a local or regional one, I mean a climate policy that has substantial ramifications for parties around the world. Decision-making on global climate policy includes a wide range of considerations: who is included in (or excluded from) relevant decision-making processes, the form taken by participation in such processes, whether the consent of those likely to be affected by some decision is secured, and so on. To address these considerations in normative terms is to take on the question of procedural justice—roughly, the question of how decisions ought to be made—in regard to global climate policy-making. Unlike distributive justice, which pertains to how harms and benefits ought to be shared across persons, procedural justice pertains to the processes whereby policy options are deliberated, evaluated, settled upon, and reviewed. It is possible for some policy to yield a distributively just outcome (at least in the axiological sense) and yet fail to be procedurally just—for example, because the decision to implement that policy was made solely by some paternalistic party, without allowing for meaningful input from parties that have a claim to offer such input.

There are serious concerns regarding whether SRM is likely to be deployed in a procedurally just fashion. If deployed on a global scale, SRM would entail substantial impacts (potentially both harmful and beneficial) and risks for various parties around the world. We may identify these as interested parties, as they would have a stake in whether (and how) SRM is deployed. Decision-making on climate policy might fail to be procedurally just by being unfair to such interested parties. Plausibly, fairness requires that each party to be substantially affected or put at non-negligible risk have an opportunity to contribute to decision-making on whether or not SRM is deployed. At any rate, it seems clear that it would be unfair to exclude an interested party from decision-making procedures regarding SRM deployment. This does not entail that merely allowing such a party the opportunity to vote on potential policies would be sufficient to make decisions about SRM deployment procedurally fair, of course. Genuine procedural fairness (and justice) may require that interested parties have the opportunity to take part in deliberations at various stages of decision-making, including decisions about what policy options are to be on the table and what research trajectories are to be pursued.[2] We might identify exclusion of at-risk parties from such deliberations as a form of *prima facie* procedural injustice. Ideally, SRM would be deployed only with the consent of such affected or at-risk parties, including members of Indigenous communities whose voices historically have not been heard on environmental matters in which they have substantial stakes.[3] As a matter of ideal theory, it might be the case that procedural justice requires that such parties consent prior to implementation of any climate policy that will substantially affect them or put them at non-negligible risk. Such policies would include, but are not limited to, SRM. However, because the preferences of various parties regarding climate policy are likely to diverge, it is unlikely that all would consent to the choice of a single response to anthropogenic climate change, whether or not it involves SRM.

Now one might object that universal consent is too demanding as a requirement for procedural justice. After all, it is unlikely that all interested parties will consent to *any* global policy, climate-related or otherwise. However, the fact that universal consent is unlikely is not relevant from the point of view of ideal theory. Ideal procedural justice holds in cases of full compliance with our duties of justice. If all parties fully complied with their duties of justice, their pursuit of self-interest would be constrained by (justice-relevant) moral considerations. It is not far-fetched to think that universal consent would be possible under those (unlikely) conditions. However, the fact that universal consent is very unlikely in our actual world is indeed relevant to non-ideal-theoretic conceptions of procedural justice. Plausibly, non-ideal procedural justice neither aims for nor requires universal consent among interested parties, and this is because such justice is constrained by political feasibility. As I will argue, one thing a non-ideal theory of procedural justice should specify is how to preserve fairness (or as much of it as possible) in decision-making when universal consent is not feasible. Nonetheless, it may be helpful to keep in mind that universal consent could be a target of ideal justice.

In this chapter, I will first argue that we have three distinct reasons for favoring procedural justice in climate decision-making. I will then provide an account of procedural fairness that can be used to ground judgments of procedural justice and injustice, thanks largely to a notion of "incomplete fairness." Finally, I will make the case that, due to tension between the political feasibility and moral permissibility conditions of non-ideal justice, it could be very difficult for SRM to be non-ideally just (in the deontic sense).

Why We Should Care About Procedural Justice

There are at least three reasons why we should favor procedural justice in decision-making about climate policy. First, moral agents have an obligation to comply with the demands of procedural justice. This obligation might be rooted in a more basic obligation to respect persons. An obvious way of failing to respect persons is to deny them an opportunity to contribute to decisions that substantially affect them, such as by unilaterally imposing very risky policies without consulting those who are put at risk. As we shall see, the parties who are put at risk are plausibly taken to have legitimate claims to contribute to such decisions. To ignore those claims outright is arguably a form of disrespect to the claimants, perhaps because doing so fails to regard them as one's moral equals. To be clear, our obligation to comply with procedural justice cannot require that all such claimants have their wishes fully satisfied. For one thing, this would often be impossible, given that many policy decisions involve various claimants with divergent interests and aims, such that satisfying one party's wishes necessarily involves thwarting (to some extent) another party's wishes. More fundamentally, it is not the business of procedural justice to secure good outcomes—although procedurally just decisions might tend to do that as well—but rather to ensure that appropriate processes are followed when it comes to deciding what to do. It is therefore possible that a procedurally just policy decision would impose serious risks on many parties, perhaps because there is no feasible option for avoiding this. In the case of climate policy, this is rather likely, given the competing preferences of various players and the imperfections of our feasible options, including realistic mitigation efforts. Nonetheless, the means whereby such decisions are made can conform to the requirements of procedural justice, and our obligation to respect other persons entails that we ought to do so. In short, the fact that we have obligations to comply with procedural justice provides us a purely moral reason to support such justice in decisions related to climate policy.

The second reason to favor procedural justice in climate policy is pragmatic, for it offers us a way to navigate both scientific and ethical uncertainty. Given Gardiner's "theoretical storm," we are unlikely to resolve the far-reaching uncertainty regarding various issues relevant to climate policy, at least not to the satisfaction of all interested parties. As we have seen in preceding chapters, reasonable positions are available on a variety of issues, including the likely geophysical and social effects of adopting any particular policy, what counts as a distributively just state of affairs, whether (and at what rate) to discount future well-being or commodities, how much compensation to provide to victims of SRM (or emissions-driven climate change), how we ought to behave in the face of deep uncertainty, and so on. Although I have defended positions on some of these questions, it is clear that disagreement will remain, particularly when it comes to questions of values and obligations. Climate change is an issue regarding which common-sense ethical norms and ways of thinking may not always be appropriate. We are not used to regarding our everyday actions as potentially involving serious harm and injustice to geographically and temporally distant persons, for example. Because of this, the ethics of climate change is likely to be less responsive to what we might call an "intuitive" approach to ethical living. In many localized, everyday cases—say, whether I should break some promise to a friend—intuition is good enough. This is not to deny that there can be hard cases. Perhaps in some situation there will be a good ethical reason for me to break a promise. Usually, however, it is intuitively clear that we ought to keep our promises, and (thankfully) we need not engage in ethical theory to determine this. But climate change is not plausibly construed in this way.

Letting procedural justice take its course is very useful here. Rather than imposing some climate policy on interested parties, one might instead let those parties decide for themselves, within a framework that includes the appropriate parties and gives appropriate weight to their input on relevant decisions. If there is ample uncertainty about (say) what distributive justice requires, this approach would defer decision-making on distributive matters to those who have a stake in the matter. Of course, even assuming impeccable adherence to the standards of procedural justice, this is no guarantee that the decision-makers will succeed in securing distributively just outcomes. They might fail in this task for various reasons. All else being equal, however, we are likely to be better off navigating ethical uncertainty via procedurally just, collective decision-making than via impositions of controversial proposals by individuals. This is because, plausibly, a procedurally just framework will allow for decision-making that is open to various points of view, which may provide a helpful check on controversial proposals. Importantly, this assumes that the involved parties are acting in good faith and complying with procedural justice. In the real world, this cannot be safely assumed. However, at present this unrealistic assumption is not problematic. For now, my claim is only that we have pragmatic reason to support procedural justice in climate decisions. I will address below the further question of how to achieve some form of procedural justice in our non-ideal world.

This argument for the pragmatic value of procedural justice immediately invites an objection, namely that the appropriate standards of procedural justice are themselves subject to substantial uncertainty. Accordingly, the theoretical storm may include not just substantive matters (e.g., regarding distributive justice) but also procedural ones. If that is the case, then it is unclear that my pragmatic appeal to procedural justice will work here. After all, if the standards of procedural justice are open to (reasonable) debate within ethical and political theory, then what standards of procedural justice should we use when it comes to making decisions that must navigate *other* controversial features of the theoretical storm? To take this objection further, it seems that, unless we arbitrarily choose some account of procedural justice, one cannot avoid wading into the theoretical controversy and making a case for some particular account of procedural justice. Once we have done that, however, we might as well wade into other theoretical controversies relevant to climate policy, in which case procedural justice arguably loses its pragmatic value.

This objection is on the right track in noting that, at least to some extent, we cannot avoid the theoretical controversy surrounding the standards of procedural justice, as we might be able to do if those standards were obvious or the object of near-universal assent. Nonetheless, this fact does not have the implications that the objector fears. Uncertainty in regard to procedural justice is but one (relatively small) aspect of the theoretical storm. We cannot avoid addressing it, for the reasons noted in the objection, but this does not entail that we must resolve other aspects of that storm in the same manner. Rather, we can do our best to work out the theoretical intricacies of what would count as a procedurally just framework for climate policy, then defer to that framework for addressing the (many more) substantive issues that are matters of ethical uncertainty. In short, there is no reason why engaging in ethical theorizing about procedural justice would require us to do the same for those other matters. Admittedly, we might fail in the task of determining what procedural justice involves here, but there is no way to avoid this possibility. If procedurally just climate decision-making has the pragmatic value I am attributing to it, then it is perfectly reasonable to defer to it when our theoretical tools are not up to the task at hand. This does not mean that we should abandon ethical theorizing in an attempt to reduce uncertainty, but simply that, so long as the uncertainty remains, we have a pragmatic reason to defer to procedurally just decisions on those uncertain issues.

A third reason to care about procedural justice is that, in the face of ethical uncertainty regarding how we ought to proceed in some matter, interested parties arguably have a right to make that decision themselves. By analogy, consider a patient suffering from a mysterious, life-threatening condition. Her doctors suggest two possible courses of action, both of them risky, and the doctors are uncertain which is the better of the two. Here the patient should be permitted to make her own choice. After all, it is uncertain which is the better option, and she is the one who will be assuming the risks of either treatment. Similarly, given uncertainty about what we ought to do as a matter of climate policy, interested parties presumably have a right to decide for themselves how to proceed. Now to take this right seriously in the case of climate policy requires honoring the standards of procedural justice. This because, unlike the patient in our analogy, those parties who have an interest in what climate policies get adopted are numerous and have variegated preferences, and therefore the right to decide how to proceed is shared by many different parties who are unlikely to agree on a single course of action. So we must consider how best to respect each party's right to decide on this matter, including who is to be party to decision-making and how much weight is to be given to each party's input. This is the domain of procedural justice. Accordingly, at least in the case of climate policies having extra-local impacts, respecting this right requires favoring procedural justice.

This third reason is distinct from the first reason to care about procedural justice, because this right of interested parties to decide for themselves is rooted in substantial uncertainty regarding what ought to be done. Because it is unclear what the appropriate course of action is, the relevant parties are entitled to decide for themselves which (likely risky) option to take. In line with the first reason given above, it is also the case that our duties of justice plausibly require us to respect procedural justice, but that reason holds even in cases that are free of uncertainty, whereas this third reason is operative only because of the uncertainty regarding what ought to be done. Likewise, this third reason is distinct from the second reason. Suppose that, contrary to what I claimed above, there is reason to believe that the collective of interested parties is less likely than some individual to navigate the relevant ethical uncertainty in an appropriate way. Arguably, it is the interested parties who have a right to decide how to proceed, for they are the ones assuming the risks no matter which option is chosen. This holds even if the individual in question is herself an interested party, because she would be but one among many such parties. To respect all right-holders in this situation would involve not prioritizing one such person above others.

So we have at least three reasons to want procedural justice in decisions of climate policy. First, independently of other considerations, we simply have an obligation to honor procedural justice, perhaps rooted in our obligation to respect persons. Second, given ethical uncertainty about how we ought to proceed, there is pragmatic value in letting the appropriate parties (perhaps via representatives, as I discuss below) decide for themselves. Third, given this same uncertainty, appropriate parties have a right to decide for themselves how to proceed, at least in cases in which all courses of action involve substantial risk. Next, I will develop an account of procedural justice that is rooted in fairness.

A Theory of Procedural Fairness

Procedural justice is plausibly taken to involve fair decision-making. While it is often obvious that certain decision-making processes involve fairness or unfairness, we need some theory of fairness in order to explain why certain cases would be procedurally fair or unfair, as well as to guide the crafting of a fairness-based account of procedural justice for climate decisions. I rely here on Broome's influential theory of fairness.[4] This theory both fits with and explains our intuitions about clear cases of fairness or unfairness, doing so in a simple and plausible fashion. I should mention that, as some of the following examples will indicate, this theory of fairness is often used to determine fair *distributions* of goods. For example, Morrow uses Broome's theory of fairness to address the question of how emissions entitlements should be distributed among different countries.[5] Nonetheless, we will see that this theory of fairness is also readily (and plausibly) applied in determining cases of procedural fairness and unfairness.

On Broome's account, fairness is the proportional satisfaction of legitimate claims that various parties have upon some good. A claim is a particular kind of reason why some person should receive some good, namely a *prima facie* duty that is owed to that person herself. Such a duty entails that this person is owed some share of the good in question. If multiple parties have claims upon some good, then fairness requires that good to be divided in proportion to the respective strength of these parties' claims. If all parties have an equally strong claim on some good, then fairness requires that good to be divided among them equally. If some party has a stronger claim on some good than another party, then fairness requires that the former receive a greater portion of that good than the latter, where the exact portion is determined by how much stronger the claim of the former is.

I will not attempt to provide necessary and sufficient conditions for what counts as a (legitimate) claim, but it is plausible to take desert to constitute one type. Let us put aside potential non-desert claims for the moment. If you deserve some divisible good that I do not deserve, then you have a claim on that good while I lack a claim on it. If no one else deserves that good—and recalling that we are ignoring potential non-desert claims for the moment— then fairness requires that good to be allotted to you in full, since you are the only party possessing a claim on it. Alternatively, if some others deserve that good as well, then fairness requires that good to be allotted to deserving parties according to how much they deserve it. Given equal desert among all parties, an equal allotment of the good would be fair. Given unequal desert among parties, the good should be allotted in proportion to that desert. For example, if you deserve the good twice as much as I deserve it, and if no other claims are relevant, then fairness requires that your share of the good be twice as great as my share.

Needs might constitute another type of claim. For example, if some person needs an organ transplant in order to continue living, we might think that this person has a claim on available organs of the required kind. If so, then this person is owed an organ. Yet if there are fewer organs in the donation bank than there are persons in need of transplants, fairness requires that the bank of organs be allotted while taking into account the extent to which each party needs a transplant. For example, a fair distribution of limited organs might involve giving transplants to those whose need is most urgent while deferring transplants for those whose need is least urgent—again, assuming that there are no other types of claim that are relevant here.

Different types of claim may be simultaneously relevant in a single case. Suppose that both desert and need constitute claims, and suppose that you have a desert-based claim on some good (but lack any other claim on it) while I have a need-based claim on that good (but lack any other claim on it). What does fairness require in this case? Since Broome holds that fairness requires satisfying claims in proportion to how strong they are, fairness involves granting all claims some degree of satisfaction, with stronger claims receiving greater satisfaction than weaker claims. Applying this to the case just mentioned, fairness would require that my need-based claim and your desert-based claim be satisfied in proportion to their respective strength. If my need-based claim is much weaker than your desert-based claim, then it would be fair for me to receive a much smaller share of the good in question while you receive a much greater share of it. If these two claims are equal, then it would be fair for each of us to receive half of the good in question. And in cases in which our claims are of equal strength but the good at issue is indivisible (e.g., the last remaining organ in the donation bank), Broome suggests it would be fair to institute a lottery, for this would give each of us an equal chance of receiving the good.

Let us say that a state of affairs is "completely fair" when all claims to some good or set of goods are satisfied in exact proportion to the strength of those claims.[6] If complete fairness holds with respect to some good, then each claimant receives a share of the good that is proportional to the strength of his or her claims relative to the claims of others. Obviously, there are many cases in the actual world where complete fairness fails to hold. One cause of this is that individuals, institutions, and governments fail to comply with their duties to parties who have claims upon various goods. Recall that some party's having a claim upon some good entails a *prima facie* duty to satisfy that claim in proportion to its strength. An individual, institution, or government that ignores or prevents such satisfaction therefore violates a duty to the claimant, who is owed some degree of satisfaction regarding her claim.[7] For example, if some community has a claim upon some natural resource (e.g., because it needs or deserves it), and this claim is ignored by some institution that harvests that resource and profits from it, then the institution has violated a *prima facie* duty to that community. This is so even if that institution also has some claim upon the natural resource, for it still would have failed to give proportional satisfaction to the claim of the community.

Let us say that a state of affairs is "incompletely fair" when some claims to some good or set of goods are satisfied, but only to an imperfect degree. This captures the thought that some scenarios involve unfairness to varying extents, or that some scenarios are clearly less unfair than others. Although a simple idea, the notion of incomplete fairness lets us make sense of this truism without ignoring the ethical deficiencies of scenarios that do not fully satisfy legitimate claims in proportion to their strength. To judge something as incompletely fair is to recognize that it exhibits both virtues and vices: on the one hand it satisfies (at least partially) some legitimate claims in proportion to their strength, but on the other hand it fails to satisfy other such claims (or fails to satisfy them in full). Incomplete fairness can encompasses a wide range. On the one hand, suppose that all claims to some good have received proportional satisfaction, save one claim that receives slightly less satisfaction than it should. On the other hand, suppose that half of all claims to some good have received proportional satisfaction while the other half of claims have received only partial satisfaction. While both scenarios are incompletely fair as a result of some claims being satisfied in each, the former is clearly more fair (or less unfair) than the latter. Given two incompletely fair scenarios, there is an ethical reason to prefer the one that is more fair (or less unfair), all else being equal.

This theory of fairness can be adapted to procedural considerations, and it can be used to explain why excluding interested parties from SRM decision-making would be unfair. First, we can view the power to make and implement a decision as a divisible good, since some interested party can have more or less influence on some decision than another party. For example, a militarily dominant state might be able to impose economic sanctions on some other state unilaterally, taking into account neither the preferences of the target state nor those of its trading partners. Second, we can hold that various interested parties can have claims upon this good, namely the power to make and implement decisions—for example, in virtue of the needs of such parties. This suggests that fairness requires such decision-making power to be distributed according to the strength of those claims. If all claimants have an equal claim to make a decision on some issue, then fairness requires the decision-making power to be divided equally among all claimants. If some claimant has a much stronger claim than another, then it would be fair for the former to have a much greater impact on the decision than the latter. And so on.

The question, of course, is what grounds a legitimate claim to decision-making on climate policy. Above, I mentioned need as a plausible example of a legitimate claim, but this works best in cases of substantive, divisible goods rather than decision-making power. For instance, persons have needs for food and water, and these needs may ground legitimate claims to certain resources, but it is less clear that persons *need* to contribute to decisions that stand to affect them. Of course, including certain parties in decision-making might have instrumental value in ensuring that the needs of those parties are considered—such as by curbing the tendency of other decision-makers to overlook those needs—but this is distinct from the question of whether one's need itself grounds a legitimate claim to decision-making. Desert-based claims are also implausible candidates here, at least if we understand desert in one of the senses discussed in chapter two (e.g., as arising from one's contribution to the social good). As with need, desert-based claims may work in the case of dividing substantive goods, such as the salary some employee is to receive relative to others. We do not usually speak of parties deserving to contribute to decision-making, however. Arguably, we need some other type of claim here.

I suggest that certain agreements can ground legitimate claims to contribute to decision-making, and it is particularly plausible to think this is the case regarding global climate policy. More specifically, I suggest the following principle: if, as part of a legitimate agreement, some party has been promised the opportunity to contribute to decision-making on some matter, then they have a legitimate claim to do so. Fairness therefore requires that this party be allowed to contribute to the relevant decision-making, and so other parties have a *prima facie* obligation not to thwart or curtail that contribution. Depending on the nature of the agreement in question, we can also determine the relative strength of various parties' claims to contribute to decisions. For instance, several parties might mutually agree to a framework in which each of them is to wield equal influence on decision-making, one in which each party has a veto on any decision, or one in which the parties are to have differential influence on decisions (e.g., because some will have greater stakes in the outcomes of the decisions than others). The non-coerced, voluntary assents of the participating parties make such agreements legitimate. Such agreements entail promises to respect the agreed-upon procedures, and so it would be *prima facie* wrong to violate those procedures. Because these promises concern how some good (i.e., decision-making power or influence) is to be divided, this is also a matter of fairness. Finally, because the good to be divided is not a substantive (e.g., material) good but rather decision-making power or influence, this is a matter of procedural (rather than substantive) fairness. Assuming that procedural justice is a matter of procedural fairness, this provides a plausible account of the former.

Understanding procedural justice in terms of legitimate agreements is particularly appropriate in the case of global climate policy, because an agreement on how to proceed in making decisions regarding global climate policy is (arguably) already entailed by the United Nations Framework Convention on Climate Change (UNFCCC), which specifies general procedures for making decisions on global climate policy. My claim is that the parties to the UNFCCC have entered into a legitimate agreement that, among other things, grounds legitimate claims on decisions of global climate policy. In effect, the parties to the UNFCCC have promised one another that each of them will be afforded an opportunity to contribute to decisions under the purview of the convention. Therefore, each party has a legitimate claim on decision-making power in such cases. Because procedural fairness requires that each party receive proportional satisfaction of its legitimate claim to contribute to decisions, and because I am understanding procedural justice in terms of procedural fairness, it would be *prima facie* unjust for one party to give other than proportional satisfaction to all such claims. In other words, the parties to the UNFCCC have obligations of procedural justice to honor one another's legitimate claims to contribute to relevant decision-making.

To be clear, my focus here is on ethical forms of legitimacy and obligation, not legal ones. Regardless of the international legal implications that the UNFCCC may carry or lack, it plausibly underwrites both claims to decision-making that are ethically legitimate and ethical obligations to honor those claims by affording them proportionate satisfaction. After all, parties to the UNFCCC (i.e., virtually every country) have mutually agreed to a process whereby the challenges of climate change are to be addressed. In effect, this is to heed Moellendorf's call that we take the norms of the UNFCCC seriously, although he does not explicitly focus on procedural justice.[8] Article seven of the UNFCCC establishes a Conference of the Parties to the convention (i.e., those countries which have assented to the convention), which "shall keep under regular review the implementation of the Convention and any related legal instruments that the Conference of the Parties may adopt, and shall make, within its mandate, the decisions necessary to promote the effective implementation of the Convention."[9] Further, the text of the convention states:

> The Conference of the Parties shall, at its first session, adopt its own rules of procedure as well as those of the subsidiary bodies established by the Convention, which shall include decision-making procedures for matters not already covered by decision-making procedures stipulated in the Convention. Such procedures may include specified majorities required for the adoption of particular decisions.[10]

Because virtually all countries in the world have voluntarily agreed to this and other procedurally -relevant measures of the convention, each country has both a claim on relevant decision-making and an obligation to honor that same claim on the part of others.

So far, I have been referring to interested parties as having legitimate claims to contribute to certain kinds of decision-making in climate policy. As my appeal to the UNFCCC already indicates, this does not mean that individual *persons* should be allowed to contribute to climate decisions. Obviously, that would be unworkable, as it might include several billion decision-makers. However, the interests of appropriate parties can be represented by legitimate surrogates, such as government representatives.[11] Presumably, procedural justice can be maintained when legitimate representatives make good faith efforts to represent the collective interests of their respective populations. In a sense, the individual persons are contributing to decisions about global climate policy, albeit via an intermediary agent. This is plausible only in the case of *legitimate* representation, however. A (nominal) representative of some population might fail to be legitimate, perhaps because the government that appointed him is authoritarian and unresponsive to the petitions of its citizens. In such a case, it is not plausible to think that procedural justice is maintained, for these citizens have no influence on the decision-making to which they have a right to contribute. Notice that this issue is distinct from the question of whether such representatives would make good faith efforts to serve the collective interests of their populations. Although it is perhaps unlikely, some (nominal) representative appointed by an authoritarian regime could strive (and even succeed) in benefiting his constituents, but he is not thereby a legitimate representative of those persons, because those persons still lack any influence on the decision-making on which they have legitimate claims.

In an ideally just world, the immediate decision-makers on global climate policy would be legitimate representatives of their respective populations, and they would all make good faith efforts to represent the respective interests (or perhaps the preferences) of those populations. They would do so within the procedural framework established by their legitimate agreements with one another under the UNFCCC, recognizing one another's legitimate claims to contribute to decision-making. Of course, in the actual world, not all government representatives will be legitimate surrogates of their respective populations, and not all representatives will make good faith efforts to serve the relevant collective interests. This holds most plausibly for appointees of authoritarian or otherwise non-democratic governments. It also holds for governments with some democratic features but which nonetheless prioritize the interests of some sections of their populations over others. For instance, a country with regular democratic elections might nonetheless be plutocratic, and its governmental representatives might tend to favor the interests of the rich. In terms of procedural justice, this could amount to an immediate decision-maker seeking to satisfy the claims of the rich to a greater extent than the claims of the poor, even when the legitimate claims of both groups are of equal strength. As I have argued, we have a moral obligation to honor the requirements of procedural justice, so we may identify cases of such illegitimacy as instances (or at least indications) of non-compliance with that obligation. It would be naïve to think that we can eliminate, or even greatly reduce, such illegitimacy in the near term. That being the case, we need some idea of how to proceed given the fact of non-compliance with our obligations of procedural justice. In other words, we need an account of non-ideal procedural justice for decision-making on climate policy.

Before continuing, I need to address a potential objection about appealing to the UNFCCC in regard to decisions about CE, SRM in particular. It is true that the text of the UNFCCC does not mention SRM. Given that the text was composed in 1992, this is not surprising, as there was little serious consideration of SRM at the time. The stated "ultimate objective" of the UNFCCC is "stabilization of greenhouse gas concentrations in the atmosphere at a level that would prevent dangerous anthropogenic interference with the climate system."[12] Because SRM would not itself stabilize atmospheric concentrations of greenhouse gases, one might argue that the UNFCCC does not pertain to SRM decision-making, and therefore that this framework does not ground agreements that are relevant to the question of procedural justice with respect to SRM. Consider two responses to this objection. First, although SRM does not by itself contribute to the "ultimate objective" of stabilizing greenhouse gas concentrations, SRM is no doubt relevant to that objective. As we shall see in chapter six, the most tenable (from the point of view of non-ideal justice) SRM policies are those that include commitments to serious mitigation of greenhouse gas emissions. Indeed, Baatz and Ott argue that our obligation to mitigate emissions has primacy, even if there is also an obligation to pursue other options (e.g., CE) to supplement mitigation.[13] Moreover, what counts as a "dangerous" atmospheric concentration is likely dependent on whether SRM is deployed. If successful, SRM may reduce many of the risks of elevated greenhouse gas concentrations. For this reason, a given concentration might be dangerous absent SRM but not dangerous (or less dangerous) with SRM. So the question of whether to deploy SRM should not be (and generally is not) considered independently of the primary aim of the UNFCCC. Second, nothing in my argument presupposes or requires that the UNFCCC itself should serve as the institutional framework that regulates SRM research and deployment. Rather, my claim is only that the mutual agreements made by the signatories of the UNFCCC plausibly entail promises that ground legitimate claims to decision-making on matters of global climate policy. Some argue that the UNFCCC is the appropriate venue for governing SRM while others deny this.[14] I do not take up that issue here, but suppose for the moment that the latter view is correct. It would nonetheless be procedurally unfair for a party to the UNFCCC to deploy SRM unilaterally. This is because, as I just noted, SRM is not tenably viewed as unrelated to the aim of stabilizing atmospheric greenhouse gases. A decision to deploy SRM will have ramifications for this aim, and so the agreements that ground legitimate claims on decision-making plausibly hold here even if the UNFCCC is not the appropriate venue for governance of SRM.

Non-Ideal Procedural Justice

There are many ways in which some party might ignore or discount the demands of (ideal) procedural justice. Most obviously, some country might pursue a global climate policy that imposes severe risks on some other interested party that has a legitimate claim on decisions regarding global climate policy, and the former country might do this without even consulting the latter. For instance, a party to the UNFCCC might simply ignore the agreement it has made with other parties to respect certain decision-making procedures, thereby ignoring other parties' legitimate claims. This would constitute a kind of promise-breaking. Alternatively, some parties to the UNFCCC might violate the requirements of procedural justice in less explicit ways— for instance, by negotiating policy proposals in bad faith. This also constitutes a kind of promise-breaking, for it fails to do what the party acting in bad faith agreed to do, namely to afford proportional satisfaction to some parties' claims on decision-making power. Unlike the first type of case of procedural justice, this second type will often be difficult to demonstrate. Nonetheless, in both cases we have some party that is failing to comply with its duties of procedural justice. To be clear, my position does not require that, in order to be procedurally just, all climate policy and decision-making must occur under the auspices of the UNFCCC, nor even that all climate policy and decision-making must involve all signatories to the UNFCCC. For one thing, not all climate policies carry risks to parties that are not included in the relevant decision-making, as in the case of a country that decides unilaterally to pursue aggressive mitigation of emissions within its own territory. Further, not all risk-carrying climate policies pose risks at the global level, as in the case of a regional coalition of countries pursuing low-leverage CDR techniques, such as afforestation, within their own region. Although, in both cases, many parties to the UNFCCC are not included in decision-making, it is not plausible to think that those parties have a legitimate claim to contribute to such decision-making. This is because the relevant promise-making entailed by the UNFCCC agreement grounds such a claim for some party only in cases in which that party has a substantial stake in the matter. In many potential cases, this condition will not be satisfied, and so the question of procedural justice will not apply.

Unfortunately, SRM is particularly prone to procedural unfairness. Given the relatively low cost estimates for deployment, and given the technical feasibility of deploying without broad participation, there is potential for a single state to deploy SRM unilaterally.[15] Obviously, a genuinely unilateral decision by some party to deploy SRM would preclude all other interested parties from contributing to the decision of whether or not SRM is to be deployed. Such interested parties have legitimate claims to contribute to a decision on whether to deploy SRM, at least in cases in which that deployment is likely to have substantial global impacts.[16] As we have seen, virtually all such parties have legitimate claims on decision-making, because they have been promised an opportunity to contribute to such decision-making as part of the agreements made under the UNFCCC. By definition, unilateral decisions to deploy SRM would fail to give proportional satisfaction to these claims. This would be a clear case of procedural unfairness, and hence of procedural injustice. One might raise a question here: could there not be a case of unilateral deployment of SRM that follows in the wake of procedurally fair decision-making? For instance, perhaps some party engages in good faith discussion with other parties who have legitimate claims on decision-making and takes their input seriously, but in the end decides to deploy SRM against the wishes of the others. Did this deploying party act in a *procedurally* unfair fashion? Two things should be noted here. First, we may distinguish unilateral SRM deployment from unilateral SRM decision-making. By definition, the latter allows for no decision-making input from other parties, even those with a legitimate claim to provide such input. In the former, the party deploys SRM without the cooperation of others. In the example just considered, it might be tempting to say that the unilaterally deploying party is not guilty of procedural injustice, because it respected the legitimate claims of other parties on decision-making even though it proceeded against the wishes of those parties. However, taking procedural justice seriously involves much more than merely listening to others. In addition, the decision-relevant input of parties with legitimate claims must play some role in constraining how we proceed. This is not to say that universal consent must be secured before we may enact some global climate policy, but the unilaterally deploying party in our example completely ignores the input of the other parties. To some extent, the deploying party does not really respect the legitimate claims of others to contribute to decisions, because their input ends up having no impact on the ultimate choice of the deployer. Accordingly, although there is some distinction between a unilateral decision and unilateral deployment, that distinction does not entail that procedural justice pertains only to the former. Now we can imagine situations in which unilateral deployment of SRM might be consistent with the demands of procedural justice—for instance, a case in which interested parties consent to a single party implementing SRM. In what follows, however, let it be understood that I am using "unilateral SRM" to refer to situations in which the decision-making itself is unilateral.

Although unilateral decision-making offers perhaps the clearest example of procedural injustice, multilateral decisions can be procedurally unjust as well. Perhaps, as Horton suggests,[17] the potential for unilateral SRM deployment is greatly exaggerated. Still, multilateral decisions to deploy SRM could exclude many legitimate claimants from decision-making and thus involve procedural unfairness. For example, Ricke et al. argue that coalitions who wish to deploy SRM would have incentives to remain small, excluding broad participation in decision-making.[18] Since different parties might have competing interests that are best served by different engineered climates, it might be difficult for large coalitions to agree on a particular SRM policy. For instance, some parties might prefer less aggressive cooling than others. Thus, a small coalition of parties might deploy SRM in a way that best serves the preferences of its members. Of course, this would exclude other parties from participating in the decision-making process, including those that have a legitimate claim to do so. But this would be a case of procedural unfairness.

From the perspective of ideal procedural justice, SRM decision-making that excludes parties with legitimate claims would be unjust. Our ideal duties of justice therefore prohibit such exclusionary decision-making, whether unilateral or multilateral. There are two things to note here, however. First, SRM decision-making need not be exclusionary. One can imagine a decision-making process on some proposed SRM policy—even one that ends up endorsing its deployment—that gives fair consideration to all legitimate claimants, affording each party an opportunity to have an influence on the ultimate decision that is proportionate to the strength of its claim to do so. This is not to say that such a procedurally fair process is likely to occur, but only to note that SRM decision-making is not necessarily unjust in the procedural sense. Second, from the mere fact that exclusionary SRM decision-making would be a violation of our obligations of ideal procedural justice, it does not follow that such exclusionary decision-making would be impermissible in all possible cases. This is because we might find ourselves in the province of non-ideal justice rather than ideal justice. For example, due to the non-compliance of others with their (ideal) duties of procedural justice (e.g., via ignoring or somehow undermining required procedures), it might be politically infeasible to proceed in ways affording complete satisfaction of all legitimate claims to decision-making in proportion to their strength. Or, because of the non-compliance of others with their (ideal) duties of some other type of justice (e.g., distributive justice), it might be that following the requirements of (ideal) procedural justice would allow some other party to act in ways that bring about substantial moral ills of some other kind. In both cases, it is (intuitively) permissible to act in certain ways that would be impermissible in the case of full compliance with ideal duties of justice. The distinction between axiological and deontic senses of justice is once again relevant here. While it would be a bad thing (i.e., unjust in the axiological sense) for some parties with relevant, legitimate claims to be excluded from decision-making, it might sometimes be permissible (i.e., just in the deontic sense) to do so.

As with distributive non-ideal justice, the foregoing holds only when it is the non-compliance of *others* that makes ideal justice infeasible or (from the point of view of justice itself) counterproductive. Recalling Gardiner's appeal to the idea of moral schizophrenia, it is not reasonable for some party to use its own non-compliance with ideal justice as an excuse to pursue non-ideal justice. For instance, if some party to the UNFCCC has consistently acted in bad faith and thus undermined procedurally just decisions, we should not be impressed by this party's argument that this non-compliance justifies less stringent, non-ideally just courses of action. Rather, the appropriate thing for this party to do is to cease undermining procedurally just decision-making. If this party continues to fail to comply with its duties of (ideal) procedural justice, however, those parties who have acted in good faith and generally complied with their obligations may be justified in pursuing non-ideal measures, particularly if the non-complying parties prove recalcitrant after attempts to persuade them to act as they ought.

Recalling our discussion from chapter two, some approach to decision-making will be non-ideally just (in the procedural sense) if it is politically feasible, likely to be effective, and morally permissible. It is important to recall that the permissibility condition includes both a proportional component and a comparative component. The former involves weighing the moral ills entailed by some approach against the moral good it achieves. The latter involves considering alternative approaches, most importantly in terms of their respective ratios of goods achieved to ills entailed. Importantly, among several approaches that fail to accord full, proportional satisfaction to all legitimate claims to decision-making, some can involve much less unfairness than others. In other words, such approaches can involve varying degrees of incomplete fairness. We must keep this in mind when considering whether some course of action satisfies the permissibility condition of non-ideal justice.

As I argued in chapter two—and also in an aforementioned paper with Morrow[19]—it might be very difficult for SRM policy-making to satisfy the feasibility and permissibility conditions of non-ideal justice simultaneously. A major reason for this is that procedural fairness and feasibility are likely to pull us in opposing directions. SRM appears politically feasible in large part because it could be pursued without broad cooperation among various parties with competing preferences. That is, SRM could be pursued by a small coalition of actors or even by a single actor. As we have seen, however, this would be procedurally unjust in the deontic sense—at least from the point of view of ideal theory—because it would be unfair. Unilateral or small-scale multilateral decisions would exclude other interested parties from decision-making, including any that have legitimate claims to contribute to such decisions. This would put pressure on the moral permissibility condition of non-ideal justice. Conversely, granting proportional satisfaction to all such claims—that is, pursuing complete procedural fairness—might help an SRM policy to satisfy the permissibility condition. This would sacrifice the central feature that made SRM appear politically feasible, namely its susceptibility to unilateral or small-scale multilateral deployment. Because SRM is likely to entail substantial global impacts relevant to the "ultimate objective" of the UNFCCC, all parties to that framework have legitimate claims to contribute to decisions on whether to deploy SRM. However, this could make it difficult to secure the broad agreement needed for a procedurally just variety of SRM, given the competing interests and preferences of the many parties involved. In other words, a completely fair approach to SRM decision-making might render SRM politically infeasible.[20]

I noted in chapter two that unilateral (or small-scale multilateral) SRM might result in distributively unjust (at least in the axiological sense) outcomes. That is a plausible concern. Members of deploying coalitions are likely to favor their own interests over those of non-coalition members. In the present chapter, the concern is distinct from distributive considerations, for unilateral deployment *need* not entail distributive justice. We can imagine a benevolent (but paternalistic) agent of unilateral SRM. Let us suppose that, due to good fortune, he succeeds in avoiding (or at least minimizing) distributive injustice and other moral ills in the wake of deployment. Nonetheless, his action would carry a substantial moral ill of the procedural variety, for many legitimate claims to contribute to decision-making would have been ignored and thus received zero satisfaction. In line with what I argued above, we have two moral reasons to care about this. First, it is a *prima facie* violation of this agent's duties of justice. It is simply wrong (in a *prima facie* sense) to ignore the legitimate claims of others on decision-making. Second, this unilateralism violates the right many parties have to contribute to decisions beset by substantial uncertainty. Although we are assuming that the benevolent climate engineer succeeds in minimizing distributive injustice and other moral ills, this was due in part to good fortune. In any realistic case, there will be uncertainty regarding the (morally relevant) effects of SRM, as well as uncertainty about how some proposed variety of SRM compares to other options. Those who have a substantial stake in such uncertain matters have a right to contribute to decisions on those matters. To return to our analogy from above, this would be like a doctor administering a risky treatment to a patient without allowing his patient any input into the decision, much less securing the patient's consent. Even if, due to good luck, the risky treatment works and the patient makes a full recovery, her right to contribute to the decision has been violated. Both these moral ills put pressure on unilateral SRM's ability to satisfy the moral permissibility condition of non-ideal justice.[21] Because unilateralism entails substantial procedural injustice (in the axiological sense), it can satisfy the proportionality component of that condition only if it secures some moral good that is proportionate to the ills imposed. That is possible, but it sets the bar rather high. Of course, even if the proportionality criterion is satisfied, the comparative question would remain. Even if the goods of unilateral SRM are proportionate to its ills, there might be some other feasible and effective option that secures a ratio of goods to ills that is better still. I return to the comparative point in chapter six.

We can reduce these procedural moral ills by pursuing broader decision-making that provides a greater degree of incomplete fairness. Although it would remain a matter of axiological procedural injustice for some parties to receive anything less than the proportionate satisfaction their legitimate claims warrant, it may be deontically just to do so in some cases. In the case of unilateralism, all legitimate claims—save one, assuming the deploying party has a claim—to decision-making receive no satisfaction whatsoever. Imagine a large coalition of states that maintains intra-coalition procedural justice, achieving complete procedural fairness by satisfying all legitimate claims within the coalition in proportion to their respective strength, and which decides to deploy SRM without consulting non-members of the coalition. This is procedurally unfair to excluded parties who have legitimate claims on decision-making, but (holding all else equal) it involves less unfairness than the case of unilateral deployment. That is, the decision has both procedural vices and procedural virtues. To be clear, that some parties with legitimate claims are excluded from decision-making is a moral ill, and it counts against the (non-ideal) permissibility of deployment. Nonetheless, the prospect for multilateral SRM to satisfy the proportionality criterion is much better than the prospect for unilateral SRM to do so because the former involves less procedural unfairness. Perhaps these are cases in which the moral goods secured by SRM are proportional to the moral ills of multilateral deployment, but not proportional to the (greater) moral ills of unilateral deployment. Unfortunately, given the incentives for deploying coalitions to remain small,[22] politically feasible varieties of SRM might carry substantial procedural unfairness, while varieties of SRM that carry little procedural unfairness may be politically infeasible. This gives us reason to suspect that, in likely scenarios, SRM will have to secure very substantial moral goods in order to offset the procedural injustice it involves.

As we saw in chapters one and two, the potential moral ills of SRM are not limited to procedural ones, but also include risks of harm and of unjust distributions. Because a (non-ideally) permissible use of SRM must secure moral goods that are proportionate to all the moral ills it carries, the bar is even higher than the previous paragraph suggests. This, of course, fits with the fairly standard view that SRM should be considered only in special cases.

Even those who think we should seriously consider using SRM in the future usually acknowledge that doing so would be appropriate only because of our failure to reduce emissions to a safe level.[23] Indeed, there may be an ethical presumption against deployment of SRM.[24] At any rate, no one thinks that deployment of SRM is desirable in its own right, although perspectives diverge on exactly how we should regard this option. Gardiner considers the idea that SRM would be a lesser evil, a topic to which I return in chapter five.[25] Alternatively, Moellendorf suggests thinking of SRM as a second-best option, rather than a lesser evil.[26] Although neither author presents his view in terms of the distinction between ideal and non-ideal justice, both their perspectives at least harmonize with the idea that SRM generally would not be permissible in cases of full compliance with our duties of justice. However, Moellendorf's view allows for the possibility that SRM might be permissible under cases of non-compliance or partial compliance with those duties. In other words, when "first-best" (regarding justice) options are not available, we are permitted to turn to "second-best" options. On the non-ideal approach I am taking in this book, this is to say that SRM could be part of a global climate policy carrying moral ills (e.g., procedural unfairness, distributive injustice, and risks of harm to innocent parties) that are proportional to the moral goods it delivers. As we have now seen, however, these goods would need to be very great in order for proportionality to hold. Further, we must not forget the comparative component of the moral permissibility condition of non-ideal justice, for under certain conditions it will be possible that SRM is not (part of) the "second-best" policy available, and so we would have reason to disfavor SRM even as a matter of non-ideal justice. I return to this issue in greater detail in chapter six.

As I have argued, decision-making on SRM can be unfair (and hence unjust) to varying degrees. As a matter of non-ideal justice, we obviously have reason to favor arrangements involving less unfairness over those involving more unfairness, all else being equal. Although it would be naïve to expect complete procedural fairness in climate decisions in the real world, it is not naïve to hope that the risk of procedural fairness can be reduced, perhaps substantially. Because of this, we should look for ways to reduce the risk of unilateral or otherwise exclusionary decisions regarding SRM deployment. One option for attempting this is for the international community to adopt (and enforce) a treaty prohibiting exclusionary decisions on SRM, such as by imposing sanctions on parties that violate this prohibition. Of course, sufficiently powerful parties may be able to ignore such a prohibition if it suits their interests. Alternatively, Horton suggests that tactics drawn from international management theory might encourage multilateralism on SRM.[27] He argues that persuasion could be used to convince a skeptical state that some SRM policy is actually in its national self-interest. If such persuasion is conducted in good faith and operates in multiple directions (e.g., with less developed countries persuading more developed ones and not just the reverse), the risk of procedural unfairness may be reduced.

Treating the power to make decisions on SRM as a divisible good, such measures could be used to increase the share of that good on behalf of parties who might otherwise be excluded from decision-making altogether or who enjoy less influence on decision-making than fairness requires. Partially satisfying some party's legitimate claim to contribute to decisions on SRM would be incompletely fair and thus ethically preferable to that claim receiving no degree of satisfaction, all else being equal. I will return to this matter in chapter six.

Given that ideal justice—in both its distributive and procedural varieties—is politically infeasible in our world, we should aim for non-ideal justice. In the past two chapters, I have attempted to lay out what this would involve, arguing that there are politically feasible options for achieving some attractive (although far from perfect) ratios of moral goods to ills. However, from the mere fact that the aims of non-ideal justice are feasible, it does not follow that relevant parties will pursue these options. In the next chapter, I turn to the question of virtue. A virtuous agent is someone who, among other things, is reliably responsive to the moral reasons she has for acting in certain ways. For instance, a just person can be counted on to act justly. Under conditions of (other parties') non-compliance with duties of ideal justice, the just person will be motivated to pursue non-ideal justice. The question is whether the (likely) agents of SRM can be expected to be virtuous, or at least to act in ways resembling the actions of virtuous agents. As we shall see, the idea of the virtuous climate engineer offers a helpful model for bringing some important ethical issues into focus.

Notes

1. Some parts of this chapter are adapted from Toby Svoboda, "Aerosol Geoengineering Deployment and Fairness," *Environmental Values* 25, no. 1 (2016): 51–68.
2. Kyle Powys Whyte, "Now This! Indigenous Sovereignty, Political Obliviousness and Governance Models for SRM Research," *Ethics, Policy & Environment* 15, no. 2 (2012): 172–87.
3. Ibid.
4. John Broome, "Fairness," *Proceedings of the Aristotelian Society* 91 (1990): 87–101.
5. David R. Morrow, "Fairness in Allocating the Global Emissions Budget," *Environmental Values* (forthcoming).
6. Arguably, complete fairness also requires that no person's rights (understood as side constraints) be violated, which is perhaps distinct from whether all claims receive proportional satisfaction. See Brad Hooker, "Fairness," *Ethical Theory and Moral Practice* 8, no. 4 (2005): 329–52. I shall not address this further, however, since my primary focus is on incomplete fairness in terms of the proportional satisfaction of claims.
7. To be more precise, preventing satisfaction of some claim violates a *prima facie* duty, but it might be permissible in certain cases. Although considerations of fairness provide moral reasons for action, there may be countervailing moral

reasons (e.g., regarding overall well-being) that override them in certain cases. Given this possibility, preventing satisfaction of some claim, while always unfair, could sometimes be morally permissible, all things considered.

8. Darrel Moellendorf, "Taking UNFCCC Norms Seriously," in *Climate Justice in a Non-Ideal World*, ed. Clare Heyward and Dominic Roser (Oxford: Oxford University Press, 2016), 104–21.

9. *United Nations Framework Convention on Climate Change*, (New York: 1992), 17.

10. Ibid., 19.

11. David R. Morrow, Robert E. Kopp, and Michael Oppenheimer, "Toward Ethical Norms and Institutions for Climate Engineering Research," *Environmental Research Letters* 4, no. 4 (2009).

12. *United Nations Framework Convention on Climate Change*, 9.

13. Christian Baatz and Konrad Ott, "Why Aggressive Mitigation Must Be Part of Any Pathway to Climate Justice," in *Climate Justice and Geoengineering: Ethics and Policy in the Atmospheric Anthropocene*, ed. Christopher Preston (Lanham: Rowman & Littlefield International, 2016).

14. Matthias Honegger, Kushini Sugathapala, and Axel Michaelowa, "Tackling Climate Change: Where Can the Generic Framework Be Located," *CCLR* 2 (2013): 125; Jesse Reynolds, "Why the UNFCCC and CBD Should Refrain from Regulating Solar Climate Engineering," in *Geoengineering Our Climate? Ethics, Politics and Governance,* ed. Jason Blackstock (New York: Routledge, forthcoming).

15. David G. Victor et al., "The Geoengineering Option a Last Resort against Global Warming?," *Foreign Affairs* 88, no. 2 (2009): 64–76.

16. Arguably, unilateral deployment of low-leverage SRM techniques (e.g., small-scale afforestation or surface albedo modification) would not be procedurally unfair, as these uses of SRM are unlikely to carry substantial impacts for those parties not included in decision-making. Once again, my focus is on global climate policies, or policies that are likely to have substantial impacts across different regions.

17. Joshua Horton, "Geoengineering and the Myth of Unilateralism: Pressures and Prospects for International Cooperation," *Stanford Journal of Law, Science & Policy* 4 (2011): 56–69.

18. Katharine L. Ricke, Juan B. Moreno-Cruz, and Ken Caldeira, "Strategic Incentives for Climate Geoengineering Coalitions to Exclude Broad Participation," *Environmental Research Letters* 8, no. 1 (2013).

19. David R. Morrow and Toby Svoboda, "Geoengineering and Non-Ideal Theory," *Public Affairs Quarterly* 30, no. 1 (2016): 85–104.

20. Although they do not address the question of moral permissibility, a similar case for tension between the effectiveness of SRM and its political feasibility is made in Scott Barrett et al., "Climate Engineering Reconsidered," *Nature Climate Change* 4, no. 7 (2014): 527–29.

21. Above, I noted another reason to care about procedural justice in decision-making on SRM, which is pragmatic in nature. This reason is not particularly relevant here, however, because ignoring it does not carry a *moral* ill. While it may be imprudent to ignore the input of others who can contribute well to decisions, it is not *prima facie* wrong or bad (in a moral sense) to do so.

22. Ricke, Moreno-Cruz, and Caldeira, "Strategic Incentives for Climate Geoengineering Coalitions to Exclude Broad Participation."

23. Paul J. Crutzen, "Albedo Enhancement by Stratospheric Sulfur Injections: A Contribution to Resolve a Policy Dilemma?," *Climatic Change* 77, nos. 3–4 (2006): 211–19.

24. Christopher J. Preston, "Re-Thinking the Unthinkable: Environmental Ethics and the Presumptive Argument against Geoengineering," *Environmental Values* 20, no. 4 (2011): 457–79.
25. Stephen M. Gardiner, "Is 'Arming the Future' with Geoengineering Really the Lesser Evil? Some Doubts about the Ethics of Intentionally Manipulating the Climate System," in *Climate Ethics,* ed. Stephen M. Gardiner et al. (Oxford: Oxford University Press, 2010), 284–312.
26. Darrel Moellendorf, *The Moral Challenge of Dangerous Climate Change: Values, Poverty, and Policy* (Cambridge: Cambridge University Press, 2014).
27. Horton, "Geoengineering and the Myth of Unilateralism: Pressures and Prospects for International Cooperation."

4 Virtues

Ethicists interested in the potential deployment of CE often investigate (1) the moral value or disvalue of possible outcomes of CE or (2) the conditions (if any) under which such deployment would be morally permissible. So far, this book has followed this convention, as the preceding chapters have addressed one or both of these issues. But we also should consider (3) the virtues and vices that different agents might exhibit in deploying CE. I claim that there are virtuous and vicious manners of deployment. I will argue in this chapter that virtue-theoretic considerations of CE are not only interesting in their own right but also relevant to (1) and (2), since virtuous agents of CE are more likely than non-virtuous ones to produce morally valuable outcomes and to deploy it in a way that is morally permissible, including the sense of moral permissibility that is relevant for non-ideal justice. This gives us ample reason to explore how certain virtues and vices might be operative in CE deployment. The idea of a virtuous agent of CE also provides a fruitful way to address the question of how to act in what I call "pessimistic scenarios," or situations in which all available courses of action carry substantial, *prima facie* ethical problems. Whether or not a pessimistic scenario constitutes a genuine moral dilemma, I will argue that the virtuous agent, being practically wise, knows how to navigate pessimistic scenarios in the best manner possible.

Unlike in other chapters, here I will mostly speak of CE in general rather than SRM techniques in particular. The reason for this is that the present chapter seeks to make a general case for the value of a virtue-theoretic approach to CE, and although I appeal to several examples, considerations of virtue and vice are in principle relevant to virtually any type of CE, as well as to climate policy more generally. This is not to deny that different types of CE pose different ethical problems and opportunities. In fact, it may be that some types are especially prone to virtuous or vicious deployment. In chapter six, I shall turn to the particular ways in which a virtue-theoretic approach can help us think about particular SRM techniques, but I first need to lay out the case for taking a virtue-theoretic approach at all.

Common Approaches to Climate Engineering Ethics

One common approach to the ethics of CE deployment is to examine the potential outcomes of different CE technologies, particularly in terms of the moral goods and ills those outcomes might involve. As we saw in chapter one, an obvious cluster of questions pertains to the harms and benefits some instance of CE might bring, as well as the likely balance of those harms and benefits. For example, CE with stratospheric aerosol injections could benefit some parties by alleviating climate-related risk (e.g., due to sea-level rise) while harming others through various side effects, such as ozone depletion or precipitation change.[1] The magnitude of CE-related harms and benefits would depend on many factors, including the type of CE technology used and the intensity, location, and duration of its deployment. Ordinarily, the benefits and harms of some CE deployment would be plausibly taken to carry moral value and disvalue, respectively.

But as discussed in chapter two, it also matters how the harms and benefits of CE are distributed, both among present parties and among the present generation and future ones. Some distributions could be unjust, such as by being unequal or failing to correlate with desert.[2] Such injustice would carry moral disvalue even apart from the magnitude of total harm that might be involved. It is worth investigating how different types of CE might compare in how their harms and benefits likely would be distributed, as well as considering how risks of injustice might be reduced, a matter to which I return in chapter six.[3]

A second common approach to the ethics of CE deployment is to consider the conditions (if any) under which deploying some CE technology would be morally permissible. This is distinct from the first approach, although no doubt there is some relation between the two. From the mere fact that some deployment of CE would be more or less harmful or unjust than other climate strategies, nothing automatically follows regarding whether or not that type of CE ought to be deployed. This axiological/deontic distinction hinges on the fact that the first approach concerns what might make CE morally good or bad, whereas the second approach concerns how we ought to act. To get clear on the permissibility of deployment, it is not enough to point out the moral goods and ills of CE, although that may be part of what is required. We also need to consider what normative principles for action would be relevant in certain cases, using these to determine whether some action or policy would be morally permissible in a given case.

Although fading in prominence,[4] one common argument by advocates of researching CE has been that it might offer the best response to a climate emergency.[5] The thought behind this argument is that some types of CE (e.g., stratospheric aerosol injections) could quickly cool the climate and thus avert some impending catastrophe due to climate change. In such cases, some type of CE may be preferable to climate strategies that are much slower to have an impact on the climate (e.g., emissions mitigation) and which therefore would be unable to avert the emergency. The reasoning behind such emergency arguments is not usually made explicit, but they can be charitably interpreted as holding that emergency deployment of CE may be morally justified and hence permissible because it is better placed than any of the alternatives to promote the best overall balance of benefits over harms or justice over injustice.

But such appeals to climate emergencies are beginning to come under criticism for various reasons.[6] For example, it is unclear both what would constitute a genuine emergency and that we would be able to detect imminent emergencies with enough confidence to justify CE deployment. More recently, CE research advocates have turned to other rationales. Perhaps CE would be morally justified as a means to "buy time" for emissions mitigation, or perhaps low-intensity CE could be used to prevent climate change from exceeding certain bounds.[7] On either argument, it might be claimed that some type of CE is justified by certain moral principles (e.g., that we have an obligation to maximize human welfare or to minimize injustice), making that type of CE morally permissible despite the risks of harm and injustice it might entail.

While they have some plausibility, Stephen Gardiner has issued an important challenge to arguments of this sort by suggesting that there may be scenarios in which no response to climate change is morally permissible. Even supposing that CE deployment would be the best (or least bad, as the case may be) option available, it might nonetheless be morally impermissible. In Gardiner's terms, CE deployment might involve a "marring evil," one that is not absolved simply because there are no better options open to us.[8] This is to say that certain climate scenarios might constitute genuine moral dilemmas in the technical sense used by moral philosophers. In a genuine moral dilemma, it is impossible to avoid moral wrongdoing. In the case of climate change, this dilemma might be self-imposed, given our collective failure to mitigate our emissions. This possibility offers another way to see the distinction between the two common approaches to CE ethics. We might have a case in which some type of CE is morally better than the alternatives, and yet—because we have a genuine dilemma—it would be morally wrong to deploy that type of CE. I address this important objection in chapter four, where I argue that although some pessimistic climate scenarios look like moral dilemmas, they are better thought of as situations calling for agent-regret. However, since this argument is yet to come, I will not assume here that there is always a permissible course of action available in pessimistic scenarios, instead limiting myself to the question of how a virtuous agent would characteristically act within them. Finally, it is not my purpose in this chapter to evaluate the merits of the arguments comprising these two common approaches. Instead, I want to argue that a third approach, a virtue-oriented one, is worth pursuing. In what follows, I will sketch what this approach would involve and show how it can help address questions arising on the first two approaches.

Virtue Theory and Climate Engineering

Little attention has been paid to the role that moral virtues and vices might play in CE deployment, although there is an interesting BA thesis by Gabriel Levine exploring virtue-ethical approaches to CE. Pak-Hang Wong has developed a Confucian perspective on CE that gives virtue an important place, and Gardiner sometimes make arguments that will be of particular interest to the virtue theorist, including the claim that SRM deployment might involve a "marring evil" on an agent's life, a matter to which I return in chapter five.[9] The Royal Society's 2009 report, *Geoengineering the Climate: Science, Governance and Uncertainty*, briefly mentions a "virtue-based" approach to ethics, which it contrasts with consequentialism and deontology, but it says very little else about virtue, and vice is not mentioned.[10] Moreover, given its contrasting of virtue-based ethics with consequentialist and deontological ones, the Royal Society report might give the impression that considerations of moral virtue with respect to CE arise only if we are already committed to a virtue-ethical normative theory.

This would be a mistake. Moral virtue can play an important role within other types of normative framework, including deontological and consequentialist varieties.[11] As a working definition, we may think of virtues as morally good character traits and vices as morally bad traits. This working definition is admittedly vague, but it is initially helpful for capturing the wide variety of traits that get counted as virtues or vices by different moral theories. Immanuel Kant, for example, takes virtue to be the strength of one's will in maintaining maxims required by the moral law.[12] While not a virtue ethicist in the sense of (say) Aristotle, virtue plays an important role in Kant's (admittedly deontological) moral theory. Indeed, Kant's *Metaphysics of Morals* includes an entire "Doctrine of Virtue," which is devoted to laying out "duties of virtue," including our duty to ourselves to cultivate morally good dispositions (i.e., virtues) as part of our "moral perfection."[13] While I cannot further pursue the place of virtue in normative theories that are not virtue-*based*, it is important to note that what I say below does not require a commitment to a virtue-based normative theory, although what I say is compatible with such a commitment. Plausibly, a wide range of virtues and vices might be operative in different types of CE deployment. The Royal Society report mentions "dilemmas" of hubris and arrogance. Presumably, the authors mean to treat hubris and arrogance as vices, although the term "vice" is not used in the report. The report also mentions the virtues of kindness and generosity in its glossary entry for "virtue-based," although it is not clear if the authors take these to be particularly relevant to CE.[14]

Before continuing, some further explanation of virtue theory in general is in order. I have said that a virtue is a morally good character trait, and virtue-theoretic moral theories are indeed focused on persons (and their traits) before actions. Nonetheless, any full theory of virtue will tell us something about right and wrong action, specifying how moral actions are connected to virtue (or vice) and offering virtue-based guidance to how we ought to act.[15] With this in mind, it is plausible to hold that virtue involves characteristically acting on the basis of certain moral reasons. The virtuous agent is not merely someone who performs morally right actions, although repeated performance of such actions may be a good indicator that the agent is virtuous. Rather, the fact that someone is virtuous in some respect entails that she *reliably* responds to the relevant moral reasons. The benevolent person is someone who takes the fact that some action would benefit others as a reason in favor of that action, and she is accordingly motivated to perform it. This does not mean that she will behave benevolently at every opportunity, of course, but it does mean that she will generally appreciate the relevance of certain moral reasons and be motivated to honor them.

I rely on a similar account of the vicious agent. A vice is a morally bad character trait, and as with a virtue this involves a certain orientation with respect to moral reasons. But whereas the virtuous agent is reliably responsive to such reasons, the vicious agent is reliably unresponsive to them. A vice is therefore not just the absence of some virtue. The merely non-virtuous agent might respond to moral reasons on occasion, but unlike the virtuous agent he does not *reliably* do so. In contrast, the vicious agent can generally be counted on to fail to respond to certain moral reasons. The greedy person, for example, is generally unmoved in cases in which there is moral reason to share his resources with others—instead, he values benefits to himself, even at the expense of much greater benefits to others. Importantly, it is not necessarily greedy merely to be motivated to pursue one's own self-interest, since doing so is compatible with being motivated to share one's own resources when that is morally appropriate. What makes one's greedy is a characteristic unresponsiveness to such reasons when they arise.

We may also distinguish between virtuous (or vicious) character traits and virtuous (or vicious) *actions*. The latter are actions that a virtuous (or vicious) agent would characteristically perform.[16] Actions and traits can come apart. It is possible for a vicious or non-virtuous agent to behave on occasion as a virtuous agent would. The greedy person might sometimes act benevolently, perhaps for a cynical reason but perhaps instead because she is feeling uncharacteristically generous. Of course, a single generous act is not sufficient to make this person non-greedy, much less in possession of the virtue of generosity. This indicates that virtuous actions are distinct from virtuous traits. Likewise, it is possible for a virtuous agent to fail to behave in some instance as a virtuous person would if acting in character. Once again, however, a single instance of non-virtuous or vicious action is not sufficient to make the agent herself non-virtuous or vicious in the relevant respect, and this again indicates a distinction between vicious action and vicious traits. I shall have frequent occasion in what follows to draw on this distinction.

I will not defend a particular theory of virtue, as this is a controversial matter and would require much attention that would distract us from the purpose of this book. Fortunately, relying on a specific theory of virtue is not necessary for my ends. I take it that there are genuine virtues and vices, understood in terms of our propensity to respond (or to fail to respond) reliably to certain types of reason. This is not a controversial view. Virtue theorists diverge on some further questions, such as the nature of our propensity to respond to reasons (e.g., whether it is a habit, a desire-like attitude, or a mode of willing), the justification of particular virtues, whether virtue contributes to the agent's flourishing, whether collective entities can be genuinely virtuous, and so on. My view is neutral on these and other theoretical questions. What I will say about the virtue of justice, for example, does not depend on whether or not it is best viewed as a habit. My account requires only that justice is a genuine virtue, understood in the broad sense that I will note below, and I do not take any position on further details about what constitutes the virtue of justice.

I will now consider (briefly) a number of character traits that are widely thought to be vices and indicate how they might be operative in CE deployment. I will then do the same for character traits widely thought to be virtues. These lists are not meant to be exhaustive, but they include traits that are particularly salient to CE, as I will attempt to illustrate.

Potential Vices in Climate Engineering

Injustice

As I shall use the term, a person with the vice of injustice lacks concern for proper distributions of harms and benefits, for the proper sharing of decision-making among interested parties (i.e., procedural justice), or both. Insofar as one is unjust, the fact that some action would result in unjust distributions of harms and benefits (e.g., by burdening some parties to an extent they do not deserve) does not motivate one to avoid performing that action, at least not very much. The unjust person in fact has moral reason to care about justice, but she is reliably unresponsive to that fact. Notably, injustice does not involve wishing ill upon certain parties, as does the vice of malice—rather, the unjust person generally lacks concern for the fact that some action or policy involves unjust distributions or procedures. Because of this, one might think that injustice is merely the absence of the virtue of justice rather than a proper vice in its own right. But this will not work, because it does not capture the *reliability* of the unjust person to be unresponsive to the relevant reasons. A person who merely lacks the virtue of justice might nonetheless act in just ways, and she might do so in response to the reasons she has as a moral agent, but she cannot be counted on to do so. At the same time, this person cannot be counted on to be *unresponsive* to these reasons. This is to say that she lacks the vice of injustice. Conversely, the unjust person can be counted on to be unresponsive to reasons of justice, and this distinguishes such a person from one who merely lacks the virtue of justice.

It is not difficult to see how the presence or absence of this vice might be relevant for CE deployment, given its potential for distributive and procedural injustice, particularly in the case of SRM varieties.[17] Presented with a range of options for dealing with climate change, the unjust person will not care (very much) if some options threaten injustice to others. She is not usually motivated to see that all interested parties are included in decision-making about whether to deploy CE nor that burdens and benefits are shared in proper ways. Compared to someone lacking this vice, an unjust person is more likely to accept SRM policies that impose disproportionate risks on those who bear no moral responsibility for climate change (e.g., low emitters and future generations) and to accept SRM decision-making frameworks that exclude interested parties. This is not to say that any SRM policy involving distributive or procedural injustice necessarily entails the vice of injustice on the part of policy-makers. As I shall argue below, it is possible for a just person to advocate a policy involving injustice.

Greed

The vice of greed involves an excessive desire for benefits to oneself, often at the expense of missed benefits or even harms for others. The greedy person is motivated to take much more of some good than is reasonably needed for some purpose, including that of living a good life. What makes this a vice is not merely the desire for benefits to oneself, but rather the tendency to desire them excessively, and even when this carries heavy costs for others. In such a case, one fails to respond to the moral reason he has to abstain from hoarding his resources or pursuing further benefits, and one who reliably does so possesses the vice of greed. Of course, in practice two or more vices might be operative simultaneously, as when an unjust and greedy person refuses to share some of his resources with those in need. One's refusal to share in this way might be overdetermined. On the one hand, he might hold fast to his resources in this case because of a desire for excessive benefits to himself, and this greed might be sufficient for the person in question not to share. On the other hand, he might hold fast to his resources because he is unresponsive to the fact that (we are assuming) he has justice-based reasons to share them with others. This vice might also be sufficient for the person in question not to share. Greed picks out the former attitude, whereas injustice picks out the latter. In principle, we can distinguish these vices, even though they come together in some cases.

Greed is presumably an important causal factor in anthropogenic climate change. Some more developed countries—or at least some agents within those countries—have apparently been driven by a very strong desire to maintain short-term economic benefits for themselves rather than pursue substantial mitigation of emissions. Such greed could (but need not) also play a role in decisions to deploy CE, for some parties might choose to pursue CE as a way of allowing us to maintain our previous (arguably greedy) lifestyles, involving heavy reliance on fossil fuels in order to achieve short-term benefits for themselves. Agents might display greed in preferring a CE policy that shifts potentially large costs and risks to future generations for the sake of preserving relatively small, short-term benefits for the present generation. An example of this might be using stratospheric aerosol injections to curb temperature increases, perhaps allowing continued use of fossil fuels at reduced risk but subjecting future persons to the termination problem, or the potential for rapid global warming as a consequence of abrupt discontinuation of aerosol injections.[18]

Short-Sightedness

As I shall understand it, the vice of short-sightedness is constituted by a failure to consider (at least some of) the non-immediate implications and impacts of some action or policy. The short-sighted person gives little thought to how her actions will impact parties that are not spatially or temporally close to her—including, as the case may be, her future self. She might be inclined to courses of action that serve the interests of those close to her but thwart the interests of those who are distant. This occurs not because the agent cares little for proper distributions (injustice) nor because the agent has an excessive desire for benefits accruing to herself (greed), but simply because the agent fails to consider the wider implications of her action. Now this could be an epistemic failing, or it could be a moral failure. In the former case, we might fail to anticipate likely effects of some action (e.g., through ignorance or insufficient imagination). In the latter case, we might fail to consider morally salient features of some case, such as how some action is likely to impact others who are spatially or temporally distant from us. We may identify the first type of short-sightedness as an intellectual vice and the second type as a moral vice. In either case, as with the clearly moral vices of injustice and greed, we can plausibly understand short-sightedness as a reliable failure to respond to certain moral reasons, in this case the (epistemic or moral, as the case may be) reasons we have to consider the non-immediate impacts of our actions. Short-sightedness is particularly worrisome in the contexts of climate change and CE, as our decisions regarding emissions, adaptation, and potential CE deployment will likely have global and long-lasting impacts. A short-sighted deployment of SRM via stratospheric aerosol injections, for example, might involve overlooking the interests of future generations.

Hubris

As the Royal Society report indicates, this vice arguably receives the most attention in discussions of CE.[19] We may understand hubris as an attitude of arrogant pride, which can manifest as a belief in one's capacity to perform feats that are in truth beyond one's abilities. The potential application to CE is apparent. It might be thought that attempting to engineer or control the climate betrays hubris, perhaps because controlling the climate transcends the abilities of even the best-intentioned and most knowledgeable human beings. As with the other vices, hubris involves a characteristic failure to respond to certain moral reasons. Assuming we have moral reason to limit our assessment of what we can achieve within realistic bounds, the hubristic person is vicious by reliably ignoring this reason. Worries about hubris are closely tied to the idea that deploying CE might involve "playing God," something that we presumably have no right to do.[20] Of course, it might be remarked that we are already displaying hubris by failing to mitigate our emissions. The thought behind this remark might be that it is arrogantly prideful to fail to mitigate while ignoring the predictable consequences of that failure, although other vices might also stand behind such behavior. If this is correct, then perhaps both our emissions and CE would involve hubris, although they may do so to varying degrees. Of course, this would not excuse the hubris involved in CE, as critics of CE are typically no less critical of the behavior driving climate change. Such critics might point out that non-hubristic responses to climate change are still available, such as substantial emissions mitigation coupled with adaptation to the environmental impacts to which we are already committed. But it is also plausible to think that different types of CE are likely to involve hubris to varying extents.[21] Finally, as we have already seen with injustice and greed, hubris might go hand in hand with other vices in the context of CE. For instance, short-sightedness might encourage hubris (and vice versa), with our failure to see implications allowing us be arrogantly prideful in thinking (incorrectly) that some use of CE will be low-risk and generally unproblematic.

Moral Corruption

Finally, moral corruption is the tendency to shirk our moral responsibilities, perhaps by relying on weak excuses to justify moral non-compliance. Gardiner plausibly suggests that climate change is "a perfect moral storm" that makes moral corruption likely. This is so for several reasons: the causes and effects of climate change are spatially and temporally disperse; there is a fragmentation of agency regarding who is driving climate change; our present institutions seem socially and politically inadequate for solving the problem; and there is uncertainty in moral theory regarding what exactly our obligations are with respect to climate change.[22] For example, parties likely to be most affected by climate change will be distant from many of us in time and space, which is a case of dispersion of cause (emissions) and effect (impacts of climate change). It is no doubt psychologically easier to ignore or discount what we owe to distant, anonymous others than to do so in the case of those who are near and known to us. Gardiner likewise worries, plausibly, that moral corruption might arise in decisions about CE.[23] Faced with prospects of dangerous climate change, we might be tempted to adopt a response that allows us to maintain emissions-intensive lifestyles, not thinking very much about the risks and costs this might impose on future generations, who cannot object since they do not yet exist. Importantly, this is not a claim that climate scientists advocating research on CE are morally corrupt. Indeed, virtually all scientists who take CE seriously hold that emissions mitigation would be a better option. The concern, rather, is that decision-makers and those influencing decision-making (including national publics) might succumb to some form of moral corruption, despite the best intentions of CE research advocates.

Moral corruption is not standardly identified as a vice in virtue-theoretic discussions, but I think it plausible to treat it as one. I have defined a vice as a propensity to be unresponsive to some moral reason. Assuming (plausibly) that we have moral reason to comply with our moral obligations, the morally corrupt person is someone who is reliably unresponsive to the fact that she has reason to so comply. Hence we have a vice. We should distinguish moral corruption as a character trait from specific instances of moral corruption. One might on occasion behave in a morally corrupt fashion without being a morally corrupt person. This traces the distinction I have drawn between vices properly speaking and vicious actions. Even virtuous agents can shirk their moral obligations, behaving on occasion as a morally corrupt person characteristically would behave. But someone with the vice of moral corruption can be counted on to fail to respond to her moral reason to comply with her obligations. However, as with any vice (or virtue), this might not be obvious to an external observer, as the morally corrupt person could *seem* to respond to such reasons without actually doing so. Kant's distinction between acting from duty and acting merely in accordance with duty is a way to understand this point. The former involves acting for the right reason, namely that the action is obligatory, whereas the latter involves performing the same action but for non-moral reasons.[24] The morally corrupt person might perform the right actions while failing to do so for the relevant moral reasons.

It is important to notice that moral corruption is equivalent to neither weakness of will nor self-deception. Rather, it is a tendency to evade one's moral obligations. In some instances, self-deception might be involved in such evasion, but one could also evade obligations with full awareness. Presumably, everyone has had the experience of sincerely believing they ought to take some action and yet failing to do so. We need not lie to ourselves when it comes to evading what we ought to do, although self-deception might offer one route for such evasion. Further, it is implausible to associate moral weakness of will with moral corruption. Imagine someone who wishes to do the right thing but is simply too weak-willed to do so. It would be odd to describe this person as *evading* her obligations, for she genuinely desires to take what she believes to be the proper course of action. That being the case, moral corruption picks out a tendency that is distinct from such weakness. Finally, we might also question whether moral corruption is rightly considered to involve a stable disposition of character, as it must in order to qualify as a proper vice. But I do not see why moral corruption cannot be a stable disposition of character. Some persons clearly have a tendency to avoid their obligations, and such a tendency seems no less a character trait in the relevant sense than (say) that of greed. Like greed, moral corruption can manifest in various ways and be associated with various psychological mechanisms. Like some morally corrupt persons, the greedy person might (or might not) deceive himself when it comes to enacting this vice through action, or he might (or might not) deceive himself in reflecting on whether or not he is a greedy person. However, regardless of the other psychological features involved, what makes this person greedy is her consistent tendency to desire excessive benefits to himself, even at the expense of grave costs to others. Likewise, what makes a person morally corrupt is one's consistent tendency to shirk one's moral obligations. Like the greedy person, there can be many explanations of why and how one is morally corrupt. Despite this diversity, however, it is hard to deny that some persons do in fact consistently evade their obligations, whether through self-deception, embracing selfishness, rationalization, and so on.

It is not difficult to imagine vicious deployments of CE. An action or policy is vicious if it is one that a vicious agent would *characteristically* pursue. CE deployment could display the vices just discussed, but I will argue below that it need not do so. In fact, CE deployment might instead involve various virtues.

Potential Virtues in Climate Engineering

Justice

A person with the virtue of justice cares about the proper distributions of harms and benefits and about the proper sharing of decision-making among interested parties. Insofar as she is just, she is motivated to preserve or promote such proper distributions and properly shared decision-making, as well as to avoid bringing about improper distributions or decision-making through her own actions. She reliably sees the fact that some action or policy would bring about unjust distributions or decision-making as a reason not to pursue that action or policy, and her character is such that she is reliably responsive to this reason. When acting in conformity with her character, someone with the virtue of justice will be disinclined to pursue types of CE that are likely to yield injustice. She will instead favor responses to climate change that maintain just (or at least less unjust) distributions and decision-making, all else being equal.

Humility

As I shall understand, the virtue of humility consists of appreciating the proper limits of our own abilities. It is opposed to the vice of hubris. The humble person acknowledges that she is subject to many forces beyond her own control. Of particular interest with respect to CE is what we might call "environmental humility," which includes appreciation of the facts that we are ignorant or uncertain about many of the workings of the Earth's climate system and that we lack the ability for fine-grained control of that system. Unlike the hubristic individual, the environmentally humble person acknowledges facts of this sort and views them as providing reasons for us to temper any enthusiasm for climate control through CE. I will nonetheless suggest momentarily that, perhaps surprisingly, an environmentally humble form of CE is possible.

Benevolence

The benevolent person genuinely values the well-being of others and is motivated to benefit others through his actions, even in cases in which this imposes a cost to oneself. Unlike the virtue of justice, benevolence pertains not to securing a proper distribution of harms and benefits among parties but rather to promoting the overall well-being of relevant parties, usually persons who are distinct from oneself. We have moral reason to promote this end, and the benevolent person reliably responds to this reason. Now one might object that this sounds more like utilitarianism than virtue theory, but it is important to remember that virtue (and vice) has a role in many normative theories, not just those that are explicitly based on virtue. For instance, we may speak of utilitarian or Kantian virtues as character traits that are valuable because they (respectively) promote welfare-enhancing actions or approximate a good will. Although I have noted that virtuous persons are reliably responsive to moral reasons, I remain agnostic as to the source of those reasons, as well as the details of how those reasons function within any normative theory. Nonetheless, practically any such theory will both take benevolence to be a virtue and allow that we have moral reasons to promote the well-being of others. Of course, competing theories will provide differing accounts of why benevolence is a virtue (and so on), but that does not threaten anything I have said here. Regardless of one's theoretical commitments, the fact that some type of CE might promote net benefits to a greater degree than the alternatives would count as a reason for the benevolent person to prefer that type of CE over the alternatives. As we saw in chapter one, this fact would not provide an all-things-considered reason to favor CE deployment, for we should care about other objectives in addition to promoting overall well-being, including objectives of justice. The *fully* virtuous agent is, among other things, someone who is good at knowing how to balance various (even competing) moral aims, seeing how to act well even when the appropriate courses of action are unclear. This requires practical wisdom.

Practical Wisdom

Identified as an intellectual virtue rather than a moral one, practical wisdom is the skill of knowing how to act well. This involves the capacity to determine what actions are (morally speaking) called for in some set of circumstances. This is done not merely by applying ready-made moral rules to specific situations—rather, the practically wise agent takes into account the morally salient features of some situation, and she is proficient in seeing what is morally called for in light of those features. Plausibly, practical wisdom has both intuitive and deliberative dimensions.[25] Sometimes the practically wise person is able to see immediately what is called for, while at other times she may employ explicit reasoning to this end. Either way, she not only understands what the virtues are in the abstract, but she also is well placed to know what being just, benevolent, or humble actually amounts to under particular conditions that happen to prevail. Practical wisdom thus provides not theoretical knowledge but rather practical knowledge—it is a type of knowing-how rather than knowing-that. Practical wisdom is a skill—or perhaps a set of skills—for being ethical, and this requires not just knowing what to do but also the ability and motivation to do it.

This account of practical wisdom is admittedly vague, but that is unavoidable, for at least two reasons. First, providing a more detailed account would require a specific theory of practical wisdom, and it would not be appropriate to spend the space developing and defending such a theory in a book on CE. Fortunately, all that is needed for my argument is the idea that there is such a thing as practical wisdom in the way just described. The use to which I put practical wisdom later in this chapter is compatible with a wide range of plausible views about the more exact nature of that capacity.

Second, and perhaps more interesting, the nature of practical wisdom is itself resistant to any determinate formulation. I take it that practical wisdom does not provide a technical decision procedure for action, which would specify determinate, prescriptive outputs (how one ought to act in particular cases) on the basis of certain inputs (the salient features of those particular cases). Instead, practical wisdom provides a kind of non-technical (arguably, both intuitive and deliberative) action-guidance. It affords, at least in part, a way of "seeing," such that the practically wise person is sensitive to the morally salient features of cases. This does not consist of imposing general normative principles on particular situations, although general rules of thumb might be useful. Rather, it involves the use of various emotional, social, and rational skills in navigating (often complex) situations. Richard Kraut puts this well, noting that, at least for Aristotle, practical wisdom "cannot be acquired solely by learning general rules. We must also acquire, through practice, those deliberative, emotional, and social skills that enable us to put our general understanding of well-being into practice in ways that are suitable to each occasion."[26] Unlike a technical decision procedure, such as that of expected utility maximization (see chapter one), it is reasonable to doubt whether we can give a satisfyingly determinate formulation of something like practical wisdom. We might think that no adequate formulation can be provided and that general discussion must remain vague to some extent, simply because practical wisdom—and particularly its intuitive dimension—is not the sort of thing that can be determinately formulated.

To be clear, I am not assuming that practical wisdom is some sort of "meta-virtue" that must underlie any virtuous action, nor that virtuous character traits require supplementation by practical wisdom in order to count as genuinely virtuous. My position is more modest: practical wisdom can help us in acting virtuously, because the practically wise person is (by definition) good at knowing what actions are called for in certain cases. It is obvious that someone can be virtuous and yet fail on occasion to act accordingly. To see this, we need only consider the evident possibility that an honest person could tell a lie on occasion. This can occur due to ignorance or confusion about (at least apparently) conflicting requirements of virtue. All else being equal, practical wisdom can help the otherwise virtuous agent in avoiding such ignorance or confusion. It is true that I do not offer much here by way of analysis of practical wisdom. Understandably, one may wish to know how this virtue actually works. This, of course, is a controversial matter, with many interesting (and conflicting) accounts in the literature.[27] However, it is neither necessary nor helpful to tie my position in this book to a particular account of practical wisdom. For one thing, it is doubtful that a book on CE is the best place to set out a well-defended view on the controversial issue of practical wisdom, given the space this would require. Perhaps more to the point, the positions I take in this chapter do not require a specific account of practical wisdom. All that is needed are the following claims: that practical wisdom is possible, that it helps us sort out apparent conflicts among the virtues, and that it can be useful in thinking through what counts as virtuous CE. Virtually all account of practical wisdom will concur with the first two claims, and I have explained above why I believe the third claim holds. Accordingly, although wading into the details of practical wisdom may be an interesting exercise in moral theory, it would not be a useful one here.

In the case of climate change, there is a wide range of morally relevant issues that we presumably should take into account, including the overall well-being of present and future generations and justice to present and future persons. But it is not immediately obvious how we ought to do so, and promoting some of these values could be in tension with promoting others. While I think there is a fact of the matter about what types of action would be virtuous (or vicious) in the context of climate change, simply having the virtues of justice, humility, and benevolence is no guarantee that one will succeed in acting well. One can reliably respond to one's moral reasons for being just, humble, and benevolent, for instance, and yet fail to act virtuously because one misjudges what those reasons call for in a specific case. This is a real danger when it comes to climate change and CE, given the many virtue-theoretic reasons that seem salient to both, not to mention the many empirical issues that are relevant. Accordingly, if we are to practice virtue in such complicated cases, we need practical wisdom. The practically wise person knows how to balance various (even competing) considerations and is capable of determining what virtue calls for in such cases. This includes knowing how to act in light of various virtue-theoretic reasons, especially when there is (at least apparent) tension among those reasons. The practically wise person who is also just and benevolent, for example, is good at knowing how to be just without compromising his benevolence and how to be benevolent without compromising his justice. Such a person's rational and affective capacities are up to the task of determining appropriate courses of action, for they are attuned to the morally relevant features of complex situations.

Virtuous Climate Engineering

We are now in a position to see how virtuous deployment of CE is possible. Consider what I will call the "virtuous climate engineer." This person has the virtues of justice, benevolence, humility, and practical wisdom. She is motivated to act virtuously, and let us suppose that she consistently succeeds in doing so. She desires that climate policy promote both a proper distribution of harms and benefits and overall human well-being. She acknowledges the limits of our understanding of the climate system and our ability to manipulate it. Crucially, she possesses the practical wisdom needed to recognize the many morally relevant features of climate change and to grant them proper consideration, all while pursuing a coherent course of action. I maintain that there are plausible scenarios in which this person, while acting in character, would favor some form of CE. Since an action or policy is virtuous if a virtuous agent would *characteristically* perform that action, the possibility of the virtuous climate engineer indicates that there can be situations in which CE deployment is virtuous. This does not mean that CE would be virtuous under *any* set of circumstances, of course, but I will now argue that there are some cases in which the virtuous agent would favor some form of CE.

There may be cases in which some type of CE better serves justice or overall human well-being than non-CE options. In such cases, the just or benevolent person has reason to be in favor of deploying that type of CE. For example, if emissions are not cut substantially and relatively soon, humanity could at some point commit itself to dangerous climate change that outstrips our adaptive capacities and thus threatens severe harm that might disproportionately (and unjustly) affect those living in poverty. Once so committed, even deep cuts in emissions would be powerless to avert these moral ills, although such cuts could avert yet more harm and injustice in the further future. In a bad scenario like this, some forms of CE could be promising, such as potentially fast-acting stratospheric aerosol injections.[28] While this technique carries its own risks of harm and injustice, as we have seen in previous chapters, these risks may be less bad than other responses to the dangerous climate change to which we are otherwise committed. If so, the benevolent and just person has reason to prefer this type of CE.

Justice and benevolence might also favor some types of CE even in cases in which we are not already committed to dangerous warming. For example, use of carbon dioxide removal techniques, such as direct air capture or bio-energy with carbon capture and storage (BECCS), might allow less developed countries to reduce poverty by relying on relatively cheap fossil fuels until a widespread transition to renewable sources of energy is achieved. Poverty reduction is an obvious good, since it increases human well-being and has the potential to ease distributive injustice. Unfortunately, wide-scale poverty reduction might be difficult without the cheap energy afforded by carbon-intensive fossil fuels, but carbon dioxide removal could allay much of the harm that would normally follow from the attendant emissions. Of course, there are potential challenges for carbon dioxide removal, such as high costs.[29] Nonetheless, justice might require that those costs be paid, perhaps by countries that have benefited from high historical emissions. Now one might wonder about this last claim and how it fits with the non-ideal-theoretic approach to justice I favor in this book. It could be objected that, although justice might require that high costs be paid under ideal circumstances, it is at least unclear that this would be required in non-ideal circumstances. In particular, it is plausible to hold that ideally just actions are not required when non-ideal circumstances render those actions ineffectual. So even if we grant that ideal justice would demand that some parties pay the high costs of carbon dioxide removal, it does not follow that justice demands this in situations likely to prevail in the near future. Whether that is so will depend on whether the technology to be deployed is likely to be effective in averting or reducing injustice, given the actual circumstances that prevail.

Both present and likely future conditions involve merely partial compliance with our duties of (ideal) justice, making them the province of non-ideal theory. But this does not undermine the relevance of the virtue of justice in non-ideal scenarios. The fact that (some) others are not complying with their duties of justice does not permit one to abandon justice—it merely changes what actions may be required, since non-ideal circumstances might render ideally just actions ineffectual. In a non-ideal scenario, the just person is still motivated to respond to her reasons to be just, and she will look for the best ways of doing so. She can and should be sensitive to changing circumstances, alert to the fact (when it is a fact) that the ideally just course of action is inappropriate thanks to the non-compliance of others. So the just person is still responsive to the same reason that holds sway in ideal circumstances, but when non-ideal circumstances prevail this responsiveness may result in actions that differ from those taken in cases of the former type. This indicates why we do not need to posit some non-ideal virtue of justice, as if justice amounted to a different character trait in ideal and non-ideal circumstances. Non-ideal theory is concerned with *actions* (and policies), not virtues. So I will continue to speak of a single virtue of justice, which virtue may favor different actions or policies in different circumstances.

But even if one grants that CE deployment can be in line with the virtues of justice and benevolence, it might be doubted that the virtue of humility could ever be operative in such deployment. After all, it might seem that attempts to control the climate are necessarily hubristic, requiring us to overstep the limits of both our understanding of the climate system and our ability to manipulate it. Yet a humble climate engineer is possible. First, a climate engineer need not attempt to *control* the climate. Historically, some proposals for planetary engineering and weather modification amounted to attempts to control parts of the Earth system, and these are aptly described as having been hubristic, but few advocates of CE research appear to take that approach at present.[30] Rather, CE is often framed as a possible way to reduce climate risk, perhaps by only curbing expected temperature increases rather than attempting to return to pre-industrial temperatures.[31] Such approaches are more aptly described as attempts to influence the climate rather than to control it. Further, it might be the case that some CE technologies (e.g., roof whitening in a particular city) are, given their nature, less prone to hubristic uses than others (e.g., annual injections of several tons of sulfate aerosols). Although this might give us more reason for concern regarding technologies of the latter sort, this does not mean that their use is destined to be hubristic, for that would depend on how the technology in question is used. Second, climate scientists are quick to acknowledge the limits of our own knowledge and abilities. For example, they recognize that there is tension among some objectives we are likely to have in deploying CE, such as moderating both temperature and sea-level rise, perhaps making it impossible for climate engineers to return the Earth to a pre-industrial climate.[32] Finally, few research advocates think that CE would be desirable in its own right. Rather, CE is often viewed as possibly the least of several evils, a means to buy time for emissions mitigation, inferior to other options that we should pursue first, and so on. This suggests not only that hubris is not a necessary accompaniment of advocating CE, but also that many of those sympathetic to CE may already avoid being hubristic, at least in certain respects.

With the foregoing in mind, we can sketch a personage that is plausibly taken to be a humble climate engineer. Imagine an agent of CE who has no illusions about controlling the climate but who believes that CE could reduce climate risks (e.g., of harm and injustice), who acknowledges the limitations of our ability to modify the climate, and who views CE as a last resort, perhaps as something that becomes reasonable to deploy only as a result of our past moral failure to reduce our emissions. At the very least, this figure does not seem to display the vice of hubris. Arguably, this figure displays the virtue of humility, for he appreciates human limitations and is unenthusiastic about deployment, even though he judges it appropriate as a last resort.

To be sure, actual climate engineers might turn out to be hubristic. For instance, they might be like the proponents of DDT critiqued in Rachel Carson's *Silent Spring*, who displayed epistemic hubris in supposing that there was nothing to be concerned about when it came to the application of this chemical.[33] Yet there is no reason why climate engineers must proceed like hubristic advocates of DDT. As I envision him or her, the virtuous climate engineer is someone who is very much concerned about the unforeseen impacts of CE, and he or she will be willing to alter or cease some CE intervention should it turn out to be more harmful than anticipated. I claimed above that non-hubristic climate engineers would aim to influence the climate rather than to control the climate. Arguably, the latter may be inherently hubristic. Nonetheless, attempting merely to influence the climate is no guarantee that attempts to do so will avoid hubris, for in doing that one might still adopt an unwarranted confidence that such influence will not yield adverse effects. So, in addition to eschewing attempts to control the climate, a non-hubristic climate engineer will take on a stance of epistemic humility, fully aware that deployment might yield negative—or even unacceptable—outcomes. The virtuous climate engineer will be prepared to respond to such outcomes if they occur, as well as to adapt the climate policy in question to new information.

Practical wisdom is perhaps the most important virtue that an agent of CE can have, given the many morally relevant issues that might be in play. Perhaps unlike the virtue of humility, I see no *prima facie* reason to worry that agents of CE might lack practical wisdom, except the possibility that practical wisdom might be a rare trait among human beings in general. More than with most human activities, we have good reason to want agents of CE to be practically wise, given the high stakes. Those without it might overlook some of the morally important considerations, such as the issues of justice regarding spatially and temporally distant parties. The practically wise climate engineer would be good at taking into account these many considerations and reasoning about how to act in light of them. She would also be on guard against moral corruption. Gardiner may be right that moral corruption is on the whole more likely in the context of climate change, but such corruption is often the result of shoddy moral reasoning (e.g., supposing that the moral non-compliance of others justifies one's own non-compliance) or selective appreciation of morally salient features (e.g., overlooking the impacts of our choices on future generations). The practically wise person is not prone to these failings, for good moral reasoning and wide-ranging appreciation of moral salience are constitutive of having practical wisdom.

The practically wise agent would also be sensitive to the many non-moral considerations that are relevant to the ethics of CE, including many salient scientific and social-political facts. Obviously, it would be unwise to decide to deploy some SRM technique without serious consideration of the relevant science, including estimates of climate sensitivity, future emissions trajectories, the probabilities of various impacts of the SRM technique in question, the fallibility of scientific investigations on these matters, and so on. Likewise, it would be unwise to ignore relevant social-political facts, such as the willingness of different populations to mitigate emissions to various degrees, the impact of competing climate policies on economic growth, the relative tractability of current government policies, the capacity of various communities to adapt, and so on. On reflection, no one is likely to think it a good idea to choose a climate policy on the basis of a single domain of information. This goes for purely moral considerations as well, which by definition would not take into account the (social) fact of non-compliance with duties of justice, thus risking recommendations that are politically infeasible and unlikely to be effective, as I argued in chapters two and three. Instead, the morally motivated assessor of some CE policy should take into account various non-moral facts as well. This is not easy to do, and it is why practical wisdom is crucial for virtuous action in the context of climate change. What (say) just action involves depends on various contingent facts, including scientific, social-political, and moral ones. No ready-made rule is likely to be sufficient in specifying, with reliable plausibility, just what we ought to do. Instead, one needs the rational and affective capacities that allow one to deliberate well on how to act, being sensitive to the relevant information in doing so, and this is what the practical wise agent is good at doing.

Why Thinking About Virtuous Climate Engineering Is Worthwhile

Reflecting on what would count as virtuous (or vicious) CE is worthwhile for several reasons. First, it might be the case that some virtue-based normative theory is true, rather than (say) a consequentialist or deontological one. If so, then how we ought to act in any situation, including whether we ought to deploy CE, will necessarily involve considerations of virtue and vice. In that case, thinking about the conditions under which CE would be virtuous or vicious is obviously helpful. However, one need not be a committed virtue ethicist in order for such reflection to be useful, given that (secondly) it can help us get clear on (1) and (2) as specified above: (1) the moral value or disvalue of possible outcomes of CE and (2) the conditions (if any) under which such deployment would be morally permissible.

As for (1), the virtuous agent is more likely than a non-virtuous or vicious agent to produce morally valuable (and minimize morally disvaluable) outcomes. For example, the person who is just and benevolent is motivated to act in ways that (respectively) promote proper distributions and overall human well-being. If she is practically wise, then the person with these virtues is also skilled at knowing how best to act to promote these goods. This gives us moral reason to want agents of CE to be virtuous. All else being equal, a climate engineer with these virtues is likely to be more reliable in realizing these moral goods and avoiding opposing moral ills than (even well-meaning) climate engineers who lack these virtues. That being the case, virtue in CE is desirable even if we are right to reject all virtue-*based* normative ethical theories. As I have noted, one need not be a committed virtue ethicist in order to recognize and esteem genuine virtues. Utilitarians, for example, should hold benevolence in high regard, for such a person is reliably motivated to act in ways that enhance overall well-being. To esteem virtue in this way does not require one even to treat virtue as a component or necessary accompaniment of right action. John Stuart Mill speaks of those who "mistake the very meaning of a standard of morals, and confound the rule of action with the motive of it." He adds:

> It is the business of ethics to tell us what are our duties, or by what test we may know them . . . [U]tilitarian moralists have gone beyond almost all others in affirming that the motive has nothing to do with the morality of the action, though much with the worth of the agent. He who saves a fellow creature from drowning does what is morally right, whether his motive be duty, or the hope of being paid for his trouble; he who betrays the friend that trusts him, is guilty of a crime, even if his object be to serve another friend to whom he is under greater obligations.[34]

On a view like this, virtuous motivation is not a constituent of right action, and in fact one's motivation is deemed irrelevant to the question of rightness. Nonetheless, it is plausible to expect that the virtuous agent will be more reliable in promoting the good, for her character is such that she is responsive to the relevant moral reasons. In Mill's case, these reasons are supposed to be utilitarian ones, but one need not be a utilitarian to think that enhanced well-being is a moral good, and so even non-utilitarians can value benevolence, given its propensity to deliver that good. The person who cares only about being paid for his trouble will often lack motivation to promote the happiness of others, but the benevolent person consistently has a motivation to do so. All else being equal, a virtuous agent is thus plausibly taken to be more likely to produce morally valuable outcomes than a non-virtuous agent.

Regarding (2), the virtuous agent is more likely than a non-virtuous or vicious agent to act in morally permissible ways, all else being equal. This is because the virtuous agent is more likely both to be motivated by the right kinds of reasons and, thanks to his practical wisdom, to understand how to act in accordance with those reasons. As before, my position does not require that virtue be part of, or otherwise be necessary to, permissible action. Mill's view is once again instructive on this point. He takes it that whether an action is right or wrong is determined by the consequences it produces (or fails to produce), not by the motives of the agent. Nonetheless, someone who is reliably motivated to act rightly is more likely to do so than someone who is not so motivated, all else being equal. The virtuous agent is reliably motivated to act rightly, because he has a propensity to act in accordance with the moral reasons that are appropriate and relevant to some case, and such appropriate and relevant reasons plausibly circumscribe a domain of permissible action in the case in question. Hence, in being generally benevolent, one will tend to act in ways deemed permissible by the Millian utilitarian, for benevolent persons will tend to promote the well-being of others and avoid acting in ways that inhibit their well-being. The same point can be made using non-consequentialist examples. A person with the virtue of justice will reliably care about procedurally just decision-making. Even assuming that the virtue of justice is not necessary for one to act in a procedurally just fashion, that virtue is valuable in decision-making contexts simply because it reliably motivates its possessor to see that procedural justice be done.

Moreover, thinking about how a virtuous agent would act in some situation—or, if we do not know, asking a virtuous agent how she would act—provides a way of determining how we ought to act in such a situation.[35] While reflecting on this is not the only way to discover whether some action would be permissible (e.g., we might also rely on normative theory for this), at the very least it offers an additional tool for thinking about how to act, and one that might prove more intuitive than other methods. We might consider, for example, whether the virtuous climate engineer would favor deployment of some type of CE in a given context—or we might ask such a figure herself. The verdict we reach on this matter could helpfully guide our own judgments on whether that type of CE ought to be deployed in the given case. I think this is more helpful than it might initially seem. Many of us are inclined to appeal to the advice of others when making difficult decisions. In doing so, it makes sense to appeal to those who are exemplary in some respect, such as those who are just or benevolent, but especially those who are practically wise. It is reasonable to expect that such persons will see more clearly what courses of action are permissible and which impermissible, precisely because they are skilled when it comes to acting morally well and, in the case of practical wisdom, knowing how to do so.

By way of contrast, consider someone with the vice of moral corruption, who is thereby prone to shirking her moral obligations. One form of moral corruption involves a kind of self-deception, convincing oneself on the basis of poor reasons that one is not really subject to certain requirements. If you like, such a morally corrupt person is "skilled" at convincing himself that he is not subject to certain requirements, excusing herself from his obligations, and this will often result in that person acting in impermissible ways without acknowledging that he is doing so. Obviously, we should not seek advice on how to act from such a person. Even if the morally corrupt person is generally aware of his tendency to make excuses for himself and makes a good faith effort not to do so, his self-deception might be so practiced and well-ingrained that he is quite bad at seeing what he ought to do. On this point, the practically wise person is exactly the reverse: she is skilled at seeing what is required and not prone to self-deception about such matters. She is thus better placed to advise others than the morally corrupt person. The same is true when we compare the practically wise person to someone who is neither morally corrupt nor practically wise. While the latter is not prone to self-deception in the service of duty-avoidance, some of us are simply not very good at seeing what sorts of actions are called for in different contexts. If we are looking for advice on how to act, as we often are, it is a reasonable move to seek it from a practically wise—or, more generally, virtuous—person.

The foregoing might give the impression that morally permissible courses of action are always possible. But what should we say about climate scenarios that might constitute moral dilemmas, or situations in which all available courses of action involve something morally impermissible? First, we should note that a candidate for such a case is either a genuine moral dilemma or merely an apparent one. If the latter, as I will argue in chapter five, then in principle there is some morally permissible course of action that is available in this scenario, although perhaps one difficult to see. It is not uncommon to experience conflict between two *prima facie* obligations. In merely apparent dilemmas there is, in principle, some solution to such conflict, perhaps because one *prima facie* obligation takes precedence over the other. In some cases, it may be difficult to determine such a solution, but I submit that the virtuous agent is well placed to discover what this difficult-to-find course of action would be, once again because of the practical wisdom that allows her to weigh various considerations and to see how best to act virtuously in difficult situations. This is not a guarantee, of course. The practically wise person can make mistakes, possibly being deceived by appearances or erring in his reasoning. Plausibly, however, the practically wise, virtuous agent has a much better shot at resolving apparent dilemmas than someone lacking these features. As we shall see in the next chapter, future climate scenarios are likely to involve at least apparent moral conflict, as feasible policies all seem to involve trade-offs among important goods. I call these "pessimistic scenarios," for within them all available courses of action are *prima facie* morally problematic.

But contrary to what I will argue in the next chapter, suppose that there is some climate scenario constituting a genuine moral dilemma, in which there is no course of action that is permissible. Here we would have a morally tragic scenario, a situation in which all possible actions involve wrongdoing. In such a genuine dilemma CE deployment would be, all things considered, morally impermissible even if it was better (or less bad) than any of the alternatives. Here there is no answer to the question of how we ought to act, for it is not the case that any of the possible actions or policies ought to be implemented. Yet the figure of the virtuous climate engineer can help us here as well. While the virtuous agent ordinarily strives to act in permissible ways and tends to succeed in doing so, this is not possible in a genuine dilemma. But the virtuous agent does not thereby lose her motivation to respond to the relevant moral reasons. She will still seek to be just, benevolent, humble, and practically wise to the extent that circumstances allow. Likewise, she will seek to avoid acting as a vicious person would, spurning actions that are unjust, greedy, hubristic, short-sighted, or morally corrupt—again, to the extent that circumstances allow.

Assuming the genuinely dilemmatic scenario, CE deployment, like any other option, would still be morally wrong. Nonetheless, the foregoing points are important. Among exclusively impermissible actions, often some will be morally better or worse than others. The virtuous agent will be motivated to act as well as possible under tragic conditions, avoiding vicious actions and performing virtuous ones to the degree that this is possible. She will be motivated to minimize suffering and injustice, for example. In a climate scenario that is a genuine moral dilemma, the virtuous agent might judge that deployment of CE, although morally wrong, is still morally preferable to the alternatives on account of its being more in line (or less out of line) with virtue. This indicates a major advantage of virtue-oriented considerations of CE compared to considerations focused primarily on the question of whether CE would be permissible. In a genuine dilemma, it is not clear that standard normative principles are much use. They specify permissible or obligatory courses of action and so arguably have no application in cases in which there is no prospect for permissible or obligatory action. Virtue-oriented considerations are not like that. Plausibly, there is still an answer to the question of how a virtuous agent would act in a genuine moral dilemma, including whether she would be in favor of deploying CE. This alone makes it promising to pursue virtue-oriented investigations of CE.

Are Virtuous Climate Engineers Unlikely?

I will close this chapter by considering an important objection to the virtue-theoretic approach to CE that I have been defending here. The general objection is that virtuous agents of CE, although possible, are unlikely to arise in relevant contexts, and thus the ruminations of this chapter are idle. Call this the Realism Objection to Virtuous Climate Engineering. This general objection can take different specific forms, one of which might be stated as follows:

> *The Realism Objection to Virtuous CE:* Like climate policy generally, decisions about whether to deploy CE will be driven by the non-moral interests (e.g., in short-term economic growth) of powerful parties, who will also determine who gets to be an agent of CE in the first place. Virtuous persons will be allowed to serve as agents of CE only if their virtue is no hindrance to the interests of powerful parties, which is very unlikely. Hence, even supposing that virtuous agents of CE are possible, there is good reason to think that the likely agents of CE will be non-virtuous, perhaps even vicious. Accordingly, at least for the purpose of guiding climate policy, it is practically useless to take a virtue-theoretic approach to CE.

There are various ways to respond to this objection. Let us first consider a response that, although attractive in some respects, will not work here.

It might be tempting to reply that thinking about virtuous CE is purely a theoretical exercise, and thus the fact that virtuous climate engineers are unlikely to arise is irrelevant. On this response, I might claim that my goal is merely to sketch what the virtuous climate engineer would look like, and nothing about this hinges on whether that figure is likely to arise in the actual world. David Estlund defends a similar view about the nature of social justice: "If a theory of social justice is offered, and it is objected, 'But you and I both know people will never do that,' I believe the right response is (as a starter), 'I never said they would.' "[36] This can be a legitimate type of response, depending on the nature of one's project. In purely ideal theory, it is a reasonable rejoinder, for regardless of what we think about the value of purely ideal theorizing, it is certainly a consistent move for the ideal theorist to make. But this response is not readily available to me, since in this book I am advocating a non-ideal approach to climate justice and arguing that CE has something to offer *given* the fact of widespread non-compliance with (ideal) duties of justice. If considerations of virtue and vice are to offer assistance in this non-ideal project, as I am claiming, then those considerations should also be sensitive to our non-ideal circumstances. Given this position, I cannot take Estlund's route in rebuffing the Realism Objection.

Second, we might accept the general pessimism of the objection but deny its conclusion. Specifically, while granting that virtuous climate engineers are unlikely to arise in the actual world, we can coherently hold that a virtue-theoretic approach to CE still has practical value. This is so because that approach, together with the fact (if it is one) that virtuous agents of CE are very unlikely, would indicate a very important ethical problem with CE, namely that any deployment thereof in the actual world is very likely to be non-virtuous or even vicious. *If* that was the case, then we would have a very strong moral reason—although not necessarily an all-things-considered reason—to oppose CE deployment in the actual world, a reason provided in part by our virtue-theoretic investigation of CE. This would, in a sense, vindicate the virtue-theoretic approach. Given that virtuous agents are more likely to act in permissible and morally valuable ways than non-virtuous agents, and supposing that the agents of CE are very likely to be non-virtuous (or vicious), there is reason to worry that the actual agents of CE could not be counted on to act in permissible and morally valuable ways. This is so because those likely agents, unlike virtuous ones, are more responsive to the interests of powerful parties than to the relevant moral reasons. In short, one could hold that the virtue-theoretic approach usefully tells us what an acceptable form of CE would look like, hard-nosed observations about reality (allegedly) tell us that CE could never actually be deployed in such a form, and thus from a virtue-theoretic perspective we have at least a *pro tanto* reason to disfavor CE deployment.

But I favor a third response to the objection, because the second is implausibly pessimistic. There are reasons to think that decisions about CE need not be captured by powerful interests. For one thing, the discourse around CE is still young, and there are opportunities for different communities to make contributions to it, including those who maintain deep ethical worries about CE. It is reasonable to hope that various norms—and, perhaps eventually, certain laws—will arise that provide some measure of protection against abuse of CE on the part of special interests. In fact, this is already occurring to some extent. For example, the Asilomar International Conference on Climate Intervention Technologies, a meeting of CE experts held in 2010, recommended several principles that should guide CE research. These included the claims that the aim of CE research should be "the collective benefit of humankind and the environment," that such research "should be conducted openly and cooperatively," and that "[p]ublic participation and consultation in research planning and oversight, assessments, and the development of decision-making mechanisms and processes must be provided to ensure consideration of the international and intergenerational implications of climate engineering."[37] Obviously, the mere promulgation of such principles is not sufficient to secure meaningful compliance with them. Further, I am not assuming that the Asilomar Principles are the right ones, much less that they would provide a complete picture of what considerations (ethical and otherwise) ought to guide CE research. Nor do I think that norm-building in regard to CE should be limited to academic experts and formal decision-makers—indeed, procedural justice would seem to require much broader participation, including that of civil society groups and the general public, as we saw in chapter three. Nonetheless, the Asilomar case provides a good example of how such an endeavor might begin.

This case also indicates that norm-building around CE has a realistic chance of being effective. The contrast with attempts at substantial emissions mitigation is instructive. A large-scale transition from fossil fuels to sources of renewable energy in the relatively short term, for example, would require massive changes in infrastructure and would have significant economic impacts, both positive and negative, on various individuals, corporations, and countries. Accordingly, convincing powerful parties to pursue substantial mitigation requires convincing them to change practices that are deeply entrenched and in which some parties are substantially invested. Accordingly, one might think that "virtuous mitigators" are unlikely to have much influence over climate policy, because certain vested interests simply would not permit that. This seems at least consistent with what can be observed about much climate policy in the past couple of decades. Yet at least some forms of CE are interestingly different on this score. Unlike with fossil fuel consumption, humanity is not currently committed to pursuing any form of CE. Although attempts to control the weather and speculation about engineering the climate have a long history,[38] serious proposals for researching CE as a potential response to anthropogenic climate change are relatively new for many people. There are many open questions about various types of CE, which are discussed at conferences, in academic publications, and in the press. Reasonable and informed parties disagree on a variety of interesting questions about CE, including ethical ones.

Admittedly, the Asilomar Principles are aspirational. They by no means guarantee that norm-building will be effective, nor that the best norms will end up being settled upon. To be as clear as possible, I am not predicting that CE governance will prove to be effective, responsible, or just. But effective, responsible, and just governance is surely possible. Asilomar and other exercises in norm-building provide *some* reason for hope that such governance might also prove politically feasible. Whether or not that comes to fruition depends crucially on what norms and governance structures (if any) are built, as well as on what form they take. The Asilomar Principles are just one example of attempts to do this. That these attempts are taken seriously by relevant communities, scientific and otherwise, is evidence that social and ethical considerations are already "in the air" around CE. As I have argued in this book, unlike with our greenhouse gas emissions, we are not yet committed to CE, so there is an opportunity for developing (ethically acceptable) norms, an opportunity that we might (or might not) squander.

In short, humanity is in a position to have a wide-ranging discussion about CE, and this provides an opportunity for building norms that favor virtuous CE and disfavor non-virtuous CE. There is no guarantee that this will occur, of course. Whether it does depends on how the discussion of CE proceeds, who is included in that discussion, how attitudes toward CE evolve, and so on. But it is reasonable to hope, perhaps cautiously, that norms in favor of justice, benevolence, humility, and practical wisdom—and against greed, injustice, hubris, short-sightedness, and moral corruption—will come to have significant influence within the CE community (broadly conceived). Should this occur, then decisions about whether to deploy CE need not be driven by the non-moral interests of powerful parties, for those interests might be opposed by influential norms. In that case, the Realism Objection might lose its force. At the very least, it would be unduly pessimistic to accept the Realism Objection at present, and this provides grounds for pursuing virtue-theoretic consideration of CE.

Notes

1. Angus J. Ferraro, Eleanor J. Highwood, and Andrew J. Charlton-Perez, "Weakened Tropical Circulation and Reduced Precipitation in Response to Geoengineering," *Environmental Research Letters* 9, no. 1 (2014); Simone Tilmes, Rolf Muller, and Ross Salawitch, "The Sensitivity of Polar Ozone Depletion to Proposed Geoengineering Schemes," *Science* 320, no. 5880 (2008): 1201–4.
2. Toby Svoboda et al., "Sulfate Aerosol Geoengineering: The Question of Justice," *Public Affairs Quarterly* 25, no. 3 (2011): 157–80.
3. Toby Svoboda, "Aerosol Geoengineering Deployment and Fairness," *Environmental Values* 25, no. 1 (2016): 51–68.
4. Joshua Horton, "The Emergency Framing of Solar Geoengineering: Time for a Different Approach," *The Anthropocene Review* 2, no. 2 (2015), 147–151.
5. Jason J. Blackstock et al., "Climate Engineering Responses to Climate Emergencies," http://arxiv.org/pdf/0907.5140.
6. Horton, "The Emergency Framing of Solar Geoengineering: Time for a Different Approach."

7. David Keith, *A Case for Climate Engineering* (Cambridge: MIT Press, 2013).
8. Stephen M. Gardiner, "Is 'Arming the Future' with Geoengineering Really the Lesser Evil? Some Doubts about the Ethics of Intentionally Manipulating the Climate System," in *Climate Ethics,* ed. Stephen M. Gardiner et al. (Oxford: Oxford University Press, 2010), 284–312.
9. Ibid., 303; Gabriel Leopold Levine, "'Has It Really Come to This?' An Assessment of Virtue Ethical Approaches to Climate Engineering," (Thesis, Yale University, 2014); Pak-Hang Wong, "Confucian Environmental Ethics, Climate Engineering, and the 'Playing God' Argument," *Zygon* 50, no. 1 (2015): 28–41.
10. John G. Shepherd et al., *Geoengineering the Climate: Science, Governance and Uncertainty* (London: Royal Society, 2009), 39.
11. Martha C. Nussbaum, "Virtue Ethics: A Misleading Category?" *The Journal of Ethics* 3, no. 3 (1999): 163–201.
12. Immanuel Kant, "The Metaphysics of Morals," in *Practical Philosophy,* ed. and trans. Mary J. Gregor (New York: Cambridge University Press, 1996), 353–603.
13. Ibid., 6: 446–47.
14. Shepherd et al., *Geoengineering the Climate: Science, Governance and Uncertainty,* 39, 78.
15. Rosalind Hursthouse, "Normative Virtue Ethics," in *How Should One Live?* ed. Roger Crisp (New York: Oxford University Press, 1996), 19–33; Rosalind Hursthouse, *On Virtue Ethics* (Oxford: Oxford University Press, 1999).
16. Hursthouse, "Normative Virtue Ethics."
17. Svoboda et al., "Sulfate Aerosol Geoengineering: The Question of Justice"; Nancy Tuana et al., "Towards Integrated Ethical and Scientific Analysis of Geoengineering: A Research Agenda," *Ethics, Policy & Environment* 15, no. 2 (2012): 136–57.
18. Andy Jones et al., "The Impact of Abrupt Suspension of Solar Radiation Management (Termination Effect) in Experiment G2 of the Geoengineering Model Intercomparison Project (GeoMIP)," *Journal of Geophysical Research: Atmospheres* 118, no. 17 (2013): 9743–52.
19. Kirsten Meyer and Christian Uhle, "Geoengineering and the Accusation of Hubris," *THESys Discussion Papers,* 2015, www.iri-thesys.org/discussion-papers/paper-pdfs/discussion-paper-2015-3-final.pdf.
20. Forrest Clingerman, "Between Babel and Pelagius: Religion, Theology, and Geoengineering," in *Engineering the Climate: The Ethics of Solar Radiation Management,* ed. Christopher Preston (Lanham: Lexington Books, 2012), 201–20; Forrest Clingerman and Kevin J. O'Brien, "Playing God: Why Religion Belongs in the Climate Engineering Debate," *Bulletin of the Atomic Scientists* 70, no. 3 (2014): 27–37.
21. Meyer and Uhle, "Geoengineering and the Accusation of Hubris."
22. Stephen M. Gardiner, "A Perfect Moral Storm: Climate Change, Intergenerational Ethics and the Problem of Moral Corruption," *Environmental Values* 15, no. 3 (2006): 397–413; Stephen M. Gardiner, *A Perfect Moral Storm: The Ethical Tragedy of Climate Change* (Oxford: Oxford University Press, 2011).
23. Gardiner, "Is 'Arming the Future' with Geoengineering Really the Lesser Evil? Some Doubts about the Ethics of Intentionally Manipulating the Climate System."
24. Immanuel Kant, "Groundwork for the Metaphysics of Morals," in *Practical Philosophy,* ed. and trans. Mary J. Gregor (New York: Cambridge University Press, 1996), 37–108.
25. Jason D. Swartwood, "Wisdom as an Expert Skill," *Ethical Theory and Moral Practice* 16, no. 3 (2013): 511–28.
26. Richard Kraut, "Aristotle's Ethics," in *The Stanford Encyclopedia of Philosophy,* ed. Edward N. Zalta (Stanford: Metaphysics Research Lab, 2014).
27. For example, see Barry Schwartz and Kenneth E. Sharpe, "Practical Wisdom: Aristotle Meets Positive Psychology," *Journal of Happiness Studies* 7, no. 3 (2006): 377–95.

28. David Keith, Edward Parson, and M. Granger Morgan, "Research on Global Sun Block Needed Now," *Nature* 463, no. 7280 (2010): 426–27.
29. National Research Council, *Climate Intervention: Carbon Dioxide Removal and Reliable Sequestration* (Washington, D.C.: National Academy of Sciences, 2015).
30. James R. Fleming, *Fixing the Sky: The Checkered History of Weather and Climate Control* (New York: Columbia University Press, 2010).
31. David Keith, *A Case for Climate Engineering* (Cambridge: MIT Press, 2013).
32. Peter J. Irvine, Ryan L. Sriver, and Klaus Keller, "Tension between Reducing Sea-Level Rise and Global Warming through Solar Radiation Management," *Nature Climate Change* 2, no. 2 (2012): 97–100.
33. Rachel Carson, *Silent Spring*(Boston: Houghton Mifflin, 2002).
34. John Stuart Mill, *Utilitarianism,* ed. George Sher (Indianapolis: Hackett, 2002), 18.
35. Rosalind Hursthouse, *On Virtue Ethics* (Oxford: Oxford University Press, 1999).
36. David Estlund, "Utopophobia," *Philosophy & Public Affairs* 42, no. 2 (2014): 114.
37. Asilomar Scientific Organizing Committee, *The Asilomar Conference Recommendations on Principles for Research into Climate Engineering Techniques* (Washington, D.C.: Climate Institute, 2010), 9.
38. Fleming, *Fixing the Sky: The Checkered History of Weather and Climate Control.*

5 Dilemmas[1]

In this chapter, I examine an argument that SRM deployment would be morally justified in a pessimistic scenario because it likely would be the best option available—or the least bad option, if no good ones are open. Here I focus on SRM rather than CE in general, because concerns about moral dilemmas are usually expressed about SRM varieties in particular. Recall that, as I am using the term, a pessimistic scenario is a situation in which all available courses of action, including all available policy choices, are *prima facie* morally problematic. This is purposely vague, for there are many ways in which an action or policy might be morally problematic, some of which I will discuss here. I will also consider an objection to the kind of argument just mentioned, namely that a pessimistic (climate) scenario would constitute a genuine moral dilemma and hence SRM deployment would not be morally justified even if it was the best (or least bad) option. Although conceiving of a pessimistic scenario as a genuine moral dilemma accounts for some ethical worries many have about SRM, it requires the very controversial claim that there are genuine moral dilemmas, and it potentially undermines moral action-guidance in pessimistic scenarios. Instead, I argue that it is better to conceive of pessimistic scenarios as situations calling for agent-regret regarding both previous moral failures that have caused such scenarios and the moral disvalue that deployment produces. This allows us coherently to hold that SRM deployment may be morally problematic even if it ought to be deployed in some situation. The key here is to distinguish between questions of permissibility and questions of value. SRM, even if very bad in some moral sense, might nonetheless be morally permissible in certain cases.

The Least of All Evils Argument and the Moral Dilemma Objection

Some proponents of SRM research argue that deployment of SRM could be justified in climate emergencies, which are plausibly viewed as pessimistic scenarios.[2] Roughly, the argument is that, in an emergency, SRM could be preferable to any other available option, and therefore deployment could be justified or permissible as "the least of all evils." I say the "least of all evils" rather than using the more common locution, "the lesser of two evils," because even in a pessimistic scenario there will usually be more than two options. An immediate concern about this argument might be its arguably dubious appeal to emergency. Once a prominent tactic in arguing for the need to research SRM, this "emergency framing" has more recently been subject to criticism. I will mention five reasons for this.

First, one might worry that the very idea of a climate emergency is too vague to offer helpful guidance on policy. What, for example, distinguishes a genuine emergency from a set of merely bad outcomes? In some cases, it might be clear enough that a certain event, such as the imminent collapse of the Greenland Ice Sheet, would constitute a genuine emergency. But this raises the second worry, namely that it can be very difficult to detect such emergencies with sufficient confidence, at least with enough lead time to allow for SRM to be effective in averting or ameliorating the event in question. Obviously, there are epistemic limits to even the best climate models, including uncertainty and deep uncertainty regarding various impacts. Third, even if we could detect an impending emergency with sufficient confidence (whatever that would be), SRM may be limited when it comes to the emergencies with which it could assist. There is a concern, for example, that averting some threshold collapse might require very aggressive cooling, which could be harmful in other respects, especially given evidence of tension between various objectives we might want to achieve via SRM.[3] Fourth, what constitutes a genuine climate emergency seems context- and valuer-dependent. An inhabitant of a small-island state might reasonably deem a relatively modest amount of sea-level rise to be an emergency, for it might threaten the very existence of the territory, culture, or state to which she belongs. Other persons might not see modest sea-level rise to be an emergency, even if they agree that it is harmful. The point is not merely that parties will disagree about whether some event counts as an emergency, but rather that some event can be a genuine emergency for some but a non-emergency for others. This makes it unclear how decision-makers should proceed in deciding whether and how to respond to context- and valuer-dependent emergencies. Fifth, as Horton argues, appeals to emergency are subject to abuse, so we should be very cautious when it comes to declaring an emergency.[4] Especially if we think that climate change fosters an environment in which moral corruption is likely, there is a concern that corrupt agents will take advantage of declared emergencies to pursue unjust or harmful agendas.[5]

Fortunately for proponents of the "least of all evils" type of reasoning in favor of SRM, such arguments need not appeal to climate emergencies, even if some proponents have done so. Consider the Least of All Evils Argument (LAEA) for SRM, which we may present as follows:

The Least of All Evils Argument

(1) In any likely pessimistic climate scenario, all available options are likely to be bad options.
(2) If we are forced to choose among exclusively bad options, then we ought to choose the all-things-considered best of those bad options.
(3) In a likely pessimistic climate scenario, the all-things-considered best option is likely to involve deployment of SRM.
(4) So in a pessimistic scenario, it is likely the case that we ought to deploy SRM.

There are five important things to notice about this argument. First, it does not depend on any appeal to a climate emergency, so it does not fall prey to the worries just expressed about emergency framing, at least not in an obvious way. Second, it does not argue in favor of an SRM-only policy. It claims only that the best option in a likely pessimistic scenario is likely to include SRM, and so SRM ought to be deployed in such a scenario, perhaps in tandem with long-term mitigation, for example. Third, the argument does not claim that the best option would involve SRM in *any* pessimistic climate scenario, but only within such scenarios that are likely to occur. Fourth, the argument claims that SRM would be part of the *all-things-considered* best option in cases of that kind. The claim is not merely that this option would be best in some respect, but rather that it would rightly be viewed to be the best option once all the relevant factors have been considered. Finally, LAEA does not maintain that just any use of SRM would be permissible in likely pessimistic scenarios, but merely that there is some use of SRM that would be permissible. Obviously, there will be some conceivable deployments of SRM that would be quite foolish, but the proponent of LAEA can accept this, because he is committed only to the view that there is at least one permissible use of SRM in pessimistic climate scenarios.

At first glance, the argument seems to have some plausibility. The crucial and controversial premise seems to be (3). That premise certainly requires disambiguation—"best" in what sense?—and thorough defense before we accept the argument, of course. Assuming, however, that we end up accepting (3), it might seem obvious that the argument succeeds. Given that a pessimistic scenario is defined as a situation in which all options are *prima facie* ethically problematic, premise (1) is hard to reject, and who would deny that we ought to take the best option in a bad situation, as urged by premise (2)? Opponents of LAEA might therefore be expected to focus any critique on premise (3). Interestingly, however, Gardiner has critiqued something like premise (2). He argues that some future climate scenarios might constitute genuine moral dilemmas, or scenarios in which it is impossible to avoid moral wrongdoing.[6] If that is the case, then deploying SRM in a pessimistic scenario would be morally impermissible even if it was the best option available. On this view, such a scenario would be a tragic situation in which *all* courses of action would be morally impermissible. On a critique like this, LAEA fails because premise (2) is taken to be false. It is alleged not to be the case that we *ought* to choose the best among exclusively bad options, for that best option is still impermissible. Call this the Moral Dilemma Objection (MDO) to SRM deployment. Because I will be focusing on this object in the present chapter, I will have little to say here about premise (3). Instead, I take up the important task of defending that premise in chapter six, where I make the case that SRM is plausibly taken to be (part of) the best option in some pessimistic scenarios that could realistically hold in the future.

Before continuing, I should note that there are different kinds of dilemma that SRM might entail. Ott argues that deployment of SRM by one generation might impose a dilemma on some future generation. For example, if SRM involves substantial harm of its own, a future generation might be forced to choose either to continue SRM (and tolerate its harm) or to cease SRM (and face the risks associated with termination). This is reasonably construed as a dilemma in some sense, but it is not clear that it is a genuine *moral* dilemma for the future generation. It is plausible to hold, as Ott does, that it is morally wrong for the deploying generation to impose such a choice on a future generation, but it is not evident that it would be impossible for the future generation to avoid moral wrongdoing in that future case. In this chapter, I am interested in the question of whether the agents of SRM deployment in a pessimistic scenario would face a genuine moral dilemma, understood as a case in which it is impossible for them to avoid wrongdoing.[7]

One might have some initial sympathy for both LAEA and MDO. On the one hand, it seems reasonable to think that, in a bad situation, we would have reason to take the all-things-considered best option available. After all, however problematic it might be, the best option is by definition better than all others. This seems to favor the thought that SRM would be permissible in such a scenario, provided that it is in fact part of the best option. On the other hand, it seems reasonable to find SRM to be morally problematic, given its potential for harm and injustice to present and future parties, as well as its potential for vicious deployment, as noted in the previous chapter. In fact, an agent's deployment of SRM might entail what Gardiner calls a "marring evil" on the life of that person, licensing "a serious negative moral assessment of that agent's life considered as a whole."[8] The worry is that there is a class of actions that are inherently marring even if one has no choice but to perform that action, and attempting to engineer the planet plausibly belongs to this class. This seems to favor the thought that SRM could be morally impermissible, regardless of whether it is better than other available options in some scenario. I shall examine LAEA and MDO in greater detail below. For now, I suggest only that it is *prima facie* plausible to treat each of these as a reasonable position.

There is an obvious tension between LAEA and MDO. One option for resolving this tension is a wholesale acceptance of one and wholesale rejection of the other.[9] However, it would be preferable instead to maintain as many of the plausible features of both LAEA and MDO as possible, given that each position seems to capture something important and plausible. In attempting this, I will argue that we have good reason to question MDO's conceptualization of pessimistic scenarios as genuine moral dilemmas, at least in the sense in which moral dilemmas are usually understood by moral philosophers. While viewing such scenarios as genuinely dilemmatic makes sense of the plausible judgment that deploying SRM would be morally problematic in some ways, that view seems incompatible with the plausible judgment that in pessimistic scenarios we ought to adopt the all-things-considered best option available. Since it is impossible to avoid moral wrongdoing in a *genuine* moral dilemma, there simply is no course of action we *ought* to adopt in such a situation, at least in a moral sense of "ought." This is an unfortunate implication for MDO. In a scenario in which all available options would (say) result in substantial net harm, but where one option is substantially less harmful than the others, and assuming all else of moral relevance is equal, it is difficult to deny that we ought to adopt that least harmful option.[10] Yet if that scenario is genuinely dilemmatic, then we must deny this, because the least harmful option still would nonetheless entail moral wrongdoing. I will argue momentarily that this serves to undermine moral action-guidance in pessimistic scenarios, just where we most need such guidance.

On the other hand, even if LAEA is sound, it would be unsatisfactory to advocate deploying SRM in a pessimistic scenario without further consideration of the many moral issues that, in addition to the question of permissibility, are relevant. Let us imagine some scenario in which SRM would be the all-things-considered best option available, and let us imagine someone who, convinced by LAEA, cavalierly advocates SRM and takes an active role in its deployment, all while washing his hands of any of the morally disvaluable or questionable features of SRM. Call this kind of person the Satisfied Climate Engineer. He might reason as follows: "A pessimistic climate scenario leaves us limited options. Because SRM is the best of those options, we ought to deploy it. Although this will result in substantial harm to various parties, we should not regard our decision to deploy SRM as morally problematic. After all, it is the best option available, and so there is nothing morally suspect about deploying it." We should worry that this kind of rationale overlooks morally salient features of both climate change and SRM deployment. In particular, it ignores the presumable moral failure that created the pessimistic scenario in the first place, it ignores the moral disvalue produced by SRM, and it ignores the moral impact that SRM deployment may have on the lives of those agents who deploy it. MDO has the advantage of addressing these matters, but only at the cost of implying that no option is permissible in the (allegedly dilemmatic) imagined scenario.

Rather than viewing pessimistic scenarios as moral dilemmas, I argue below that it is more plausible to view them as cases in which agent-regret is appropriate. It can be the case that some action is morally permissible (even obligatory) and yet calls for agent-regret on the part of whoever performs that action. Assuming that it is part of the best option in some bad situation, it is at least *coherent* to hold both that we ought to deploy SRM and that it is morally good to harbor agent-regret for doing so. As I shall argue, an agent-regret account allows us to hold on to the thought behind LAEA that SRM might be permissible in conceivable cases, yet it also allows us to assess the Satisfied Climate Engineer's posture as morally problematic, for he lacks an attitude of regret that it would be appropriate for him to have. Before making this case, however, I first suggest some reasons to be skeptical that pessimistic scenarios would involve genuine moral dilemmas.

Some Costs of Genuine Moral Dilemmas in General

There are reasons to doubt that genuine moral dilemmas occur at all, given that they would seem to conflict with some very plausible principles. Perhaps most obviously, the principle of "ought implies can" holds that if a moral agent is morally obligated to perform some action (or to abstain from doing so), then it must be possible for that moral agent to perform that action (or to abstain from doing so). A genuine moral dilemma would be inconsistent with "ought implies can," because it is by definition a scenario in which it is impossible to do as one ought. If some climate scenario would count as a genuine moral dilemma, in the sense that all available options (including doing nothing) would be morally impermissible, then it would be impossible for a moral agent to act as she ought in that situation. Now if "ought implies can" is a conceptual truth, then such scenarios cannot hold.[11] In order to preserve this principle, it must always be possible for an agent to do what she ought to do, no matter how grim the situation. In a pessimistic scenario then, assuming this moral principle holds, at least one option must always be morally permissible.

The obvious response to this is to deny that "ought implies can" is a conceptual (or even a necessary) truth, but one must be careful here. It would be very implausible to deny this principle altogether. Most will want to hold on to "ought implies can" in some form. At the very least, this principle is attractive because it provides a powerful explanation of why we are almost always able to act as we ought. Without such a principle, this fact would merely be a happy and implausible accident. Those who are sympathetic to the idea of genuine moral dilemmas are better served to view "ought implies can" as a principle that generally holds but admits of rare exception in special cases. A good example of this approach is to hold that there are genuine *self-imposed* moral dilemmas. Arguably, this variety of dilemma does not threaten the plausible thought behind "ought implies can," although the existence of such a genuine dilemma would disqualify that principle as a conceptual (or necessary, or even universal) truth. Perhaps "ought implies can" holds in almost all cases, but not those in which an agent's own wrongdoing has created moral conflict. The standard example of this is making incompatible promises.[12] If I promise George that I will watch a movie with him at a specific time, and if I then promise Terrence that I will go for a walk with him at that same time, it is not possible for me to keep both promises. Presumably, I am obligated to keep each promise, and therefore I am obligated to keep both promises. But this is impossible, since doing so would require me to be in two different places at once. Instead, I must break at least one of my promises, and arguably this would be morally wrong to do. If so, then this would be a case of a genuine moral dilemma.

How should we view this case? We might view breaking either of my promises as morally wrong, even though it is impossible to avoid this. In that case, the situation would be genuinely dilemmatic. This would be inconsistent with "ought implies can," for once I have made the incompatible promises there is no way for me to avoid breaking at least one of them, and so I cannot do what I ought to do. But we might find this philosophically untroubling, given that the dilemma was created through my own poor decision-making in making incompatible promises. In fact, we might think that my doing so was itself morally impermissible, and in a way that is straightforwardly consonant with "ought implies can," for I could have avoided making incompatible promises, but I nonetheless made them, and I acted wrongly in doing so (e.g., at the point at which I made the second promise). Due to my own wrongdoing, I have now created a dilemma for myself, which I cannot escape without yet more wrongdoing (e.g., by breaking one of my promises). If previously I had acted as I ought to have done, I would not have made incompatible promises in the first place. But since I did make incompatible promises, I have only myself to blame for the existence of this genuine dilemma, which now forces me to act as I ought not to act.

Dilemmas of this self-imposed variety are at least less mysterious than other varieties. For example, we might be suspicious of the claim that there are "externally" imposed dilemmas, or scenarios in which an agent is forced into moral wrongdoing through no fault of her own, such as through the impermissible actions of others. *This* kind of exception to "ought implies can" seems particularly implausible. In contrast, self-imposed moral dilemmas arguably do not violate the spirit behind "ought implies can." If an agent acts as she ought to act, then arguably self-imposed moral dilemmas will not arise. It is only when an agent acts wrongly, we might think, that self-imposed dilemmas are created, and so it is within one's power to prevent such dilemmas from arising in the first place.

Initially, this seems like a promising route for proponents of MDO to take, as it is tempting to view pessimistic climate scenarios as self-imposed moral dilemmas. I assume that some parties have a moral obligation to reduce greenhouse gas emissions. If we do not fulfill this obligation, we reasonably can expect to reach a point in the future at which we are committed to dangerous climate change—if we have not done so already—including threshold collapses in the climate system, the impacts of which could be severely harmful while also outstripping our adaptive capacities. This future point is reasonably construed as a pessimistic scenario, for all courses of action in that situation are, plausibly, *prima facie* ethically problematic: for example, emissions mitigation cannot ameliorate committed climate change, adaptation is limited in terms of what goods it can protect, carbon dioxide removal may not be scalable to the extent needed, and SRM has the various drawbacks we have reviewed in previous chapters. To say that such responses to dangerous climate change are *prima facie* ethically problematic is merely to point out that there are reasonable ethical concerns about each—regarding risks of injustice, for example—and does not commit one to the view that any or all of these responses would be morally impermissible. Yet one might further view such a scenario as a self-imposed moral dilemma, since it arises from our past moral failure to reduce our emissions. Proponents of MDO might then reason as follows. While "ought implies can" is a reasonable moral principle that holds in most cases, exceptions to it can arise due to our own moral wrongdoing in previous, non-dilemmatic cases. In failing to make obligatory cuts to our emissions, we risk imposing a future dilemma on ourselves. In a bad situation of our own making—or at any rate, one that we could have avoided—we might no longer be able to act as we ought to act, but this merely reflects our past decision not to comply with our moral obligations. So SRM might be impermissible in the pessimistic scenario just mentioned, even if it was better (all things considered) than the other responses.

Let us suppose for the sake of argument that there are genuine moral dilemmas of the self-imposed variety. Having granted this, it is unclear that a pessimistic climate scenario would meet the conditions necessary for it to be plausibly considered an instance of such a dilemma. Up to this point, I intentionally have been vague concerning what agents would be involved in some putatively dilemmatic situation. Yet a scenario counts as a self-imposed moral dilemma only if the agent who faces it is identical to the agent who imposed it, for otherwise it would not be *self*-imposed. This works in the case of incompatible promises: I am the agent who (wrongly) makes the promises to both George and Terrence, and I am also the agent who must break at least one of those promises. But there is reason to doubt that a pessimistic climate scenario would be like this, for the class of agents acting in the scenario is unlikely to fall fully within the class of parties culpable for that scenario itself, even assuming that we rightly identify those who are culpable for their emissions.[13]

To begin with, these two classes can be temporally distinct from each other. For example, the wrongful emissions of earlier generations could ensure some threshold event (e.g., the collapse of a major ice sheet) that, due to inertia in the climate system, would occur only at some point in the future. In such a case, a future generation might face the decision of whether or not to deploy SRM to avert this otherwise imminent threshold collapse. But if this would be a genuine moral dilemma for the future generation, it is not a self-imposed one, since those who face the putative dilemma are not identical to those who caused it. This is relevant here because very few proponents of SRM research argue for near-term deployment of SRM.[14] Rather, such proponents tend to argue that research is needed now because deployment might be needed by future persons.[15] Given that future persons cannot be responsible (morally or causally) for emissions that occurred before they existed, and because the scenario we are imagining would be driven by emissions of this type, that scenario is not reasonably viewed as a *self-imposed* moral dilemma. Even putting questions of temporal distinctness to one side, there still could be cases of non-identity between the culpable agents of some pessimistic climate scenario and those who must respond to it. The very existence of some small-island states is threated by sea-level rise. Given continued emissions, and faced with such a harmful prospect, one such state might choose to deploy SRM in an attempt to stabilize the climate and preserve its own territory.[16] Given the low per capita and overall emissions of small-island states, it is not plausible to treat them (nor their citizens) as culpable agents of the scenario that threatens such substantial harm to them. Once again, if this would be a moral dilemma for the would-be deployers of SRM, it is not plausibly viewed as a self-imposed one.

Finally, there could be cases of partial but incomplete overlap between the set of agents culpable for a pessimistic scenario and the set of agents responding to it. That is, those who face this scenario could be "mixed," including some who were obligated to reduce their emissions but failed to do so, as well as some (perhaps many) who were not so obligated. This makes it more difficult to evaluate whether the situation could be a self-imposed moral dilemma. Describing deployment of SRM by a mixed set of agents as morally wrong because of the past wrongdoing of only some members of that set seems implausible. Why should the actions of those who are not culpable for the pessimistic scenario be morally tainted by the previous wrongdoing of their collaborators? Now one might reply that *only* agents culpable for the pessimistic scenario would act wrongly in deploying SRM, whereas their collaborators would not act wrongly in joining them. This is an interesting suggestion, but even if some individuals act impermissibly in collaborating on deployment, it does not follow that the *collective* action of deployment is impermissible. The core thought of MDO is that SRM deployment itself would be impermissible, not merely that some individuals would act wrongly in serving as agents of deployment.

Evidently, the only plausible case of a self-imposed moral dilemma of the relevant type would be a non-mixed scenario, in which the agents of SRM are themselves culpable for the pessimistic scenario. A case like this would be reasonably analogous to that of the person who makes incompatible promises. Just as the latter has himself to blame for creating a situation in which he cannot avoid breaking some promise, so these agents of SRM would have themselves to blame for the pessimistic scenario in which (let us assume) SRM is the all-things-considered best option. But the analogy is limited. Whereas the maker of incompatible promises has *only* himself to blame for the putative dilemma, the agents of SRM will constitute only some of those who are culpable for the pessimistic scenario—unless this class of agents is entirely coextensive with the class of culpable emitters, which is very unlikely and perhaps impossible (because some culpable emitters may no longer be alive, for example). Unlike the incompatible promise-maker, these agents of SRM do not have *only* themselves to blame for the scenario. For this reason, it is unclear that they would have properly imposed this scenario on themselves, rather than merely contributing to a process that caused it. In fact, depending on the scale of their culpable emissions, it might be that the pessimistic scenario would have occurred even had the agents of SRM (counter to fact) not emitted more than was permissible. They might go on to argue that, given this counterfactual, the pessimistic scenario in question is not rightly viewed as a self-imposed scenario. If that is correct, then of course the scenario cannot be a self-imposed moral dilemma either. In that case, the proponent of MDO would have to fall back on the controversial claim that a genuine moral dilemma can arise through no fault of the parties to whom it applies. As we have seen, however, this would violate "ought implies can" in a much more problematic way than would self-imposed dilemmas.

Although I have noted here some reasons for skepticism about genuine moral dilemmas in general, I obviously have not demonstrated that such dilemmas are impossible. I should stress that my argument in this chapter is *not* dependent on dismissing the possibility of genuine moral dilemmas and then straightforwardly inferring that pessimistic climate scenarios must not be genuinely dilemmatic. Rather, as we shall see, my argument is comparative in nature. The question is whether to view certain types of climate scenarios (e.g., those in which SRM might plausibly take place) as genuine moral dilemmas. The main rationale for doing so is that a moral dilemma framing captures the moral disvalue of actions performed in such scenarios. If this was the only way to capture the moral disvalue of SRM, then the costs of violating "ought implies can"—and, as I argue in the next section, undermining moral action-guidance—may be well worth paying. But I will show that this is not the only way to take seriously the moral disvalue of SRM. Conversely, the agent-regret framing that I present below lets us seriously acknowledge the moral disvalue of SRM without incurring these heavy costs. This gives us good reason to prefer the agent-regret framing of pessimistic climate scenarios, but one need not reject the possibility of moral dilemmas in general in order to accept this comparative argument.

Action-Guidance in Pessimistic Scenarios

Putting aside the concerns just mentioned, the view that pessimistic climate scenarios are genuinely dilemmatic has the further problem that it leaves us with little prospect for moral action-guidance in such cases. A genuine moral dilemma is a situation in which no available course of action is morally permissible and thus a situation in which it is impossible for an agent to avoid moral wrongdoing. It would be very odd, and perhaps incoherent, to hold that we (morally) ought to take a (morally) impermissible course of action. If all available courses of action are morally wrong, then arguably it is not the case that any of them ought to be adopted, at least in a moral sense of "ought." Therefore, in a genuine moral dilemma, there seems to be no answer to the question of how we should act. For if there was an answer to this question, then there would be some action we ought to take, and an action we ought to take must be a permissible one. But there are no permissible actions in a genuine moral dilemma, and so moral considerations seem unable to provide any kind of guidance in deciding how to act in such a dilemma.

This is at least a serious drawback for MDO, since it would be valuable to know how we ought to act when in such bad circumstances. Further, in a pessimistic scenario, there could be many feasible courses of action that differ markedly in their prospects for causing harm and injustice. For instance, it seems that in such a scenario we would have moral reason to prefer some less harmful or unjust option over some more harmful or unjust option, all else being equal. Indeed, recognition that some options are "lesser evils" seems implicitly to grant this. Yet it is difficult to see how we could make sense of this in a genuine moral dilemma. If all available courses of action are morally wrong, then we seem to be prohibited from taking any of those courses. An implication of this is that, contrary to our intuitions, we would lack *sufficiently* good moral reason to adopt substantially less harmful or unjust options over those that are more harmful or unjust, even assuming all else is equal among those options. Treating a pessimistic climate scenario as a genuine moral dilemma therefore seems to undercut potentially helpful action-guidance we might have received from moral theory. It leaves us at a loss regarding how we ought to act in such cases. This is unfortunate, for it would seem that it is in pessimistic scenarios that moral action-guidance is most needed.

One advantage of LAEA—or, more properly, the perspective it affords—is that it allows for moral action-guidance in pessimistic scenarios. Given a situation in which all our options are bad ones, LAEA directs us to choose the least bad option. The argument contends that, in likely pessimistic climate scenarios, this least bad option likely would involve SRM. At least in principle, we can determine whether or not SRM really would be included in the least bad choice. While virtually no proponent of SRM research thinks that deployment of SRM is a desirable goal in its own right, many such proponents are sympathetic to something resembling LAEA, particularly in an emergency scenario—although, as we have seen, the emergency framing is not necessary to LAEA-style arguments for SRM. From the point of view of LAEA, moral considerations, including those drawn from moral theory, can provide action-guidance in pessimistic scenarios. Viewed as non-dilemmatic cases, we can preserve the intuition that some courses of action are morally preferable to others in pessimistic scenarios.

Importantly, thinking that SRM is morally preferable to non-SRM alternatives in certain cases is compatible with thinking that SRM is also morally problematic. A Tentative Climate Engineer could reason as follows: "Deployment of SRM is morally problematic in various ways. It could cause substantial harm and injustice to present and future persons. We instead ought to make substantial cuts in our emissions in order to avoid dangerous climate change in the future. Ideally, if we reduce emissions as we ought, there would be no need for SRM in the future. But if we continue to fail in cutting our emissions to the extent required, our situation may need to be reassessed. In a pessimistic climate scenario, it could be too late for emissions cuts to avoid dangerous climate change, and adaptive measures may be insufficient to prepare for it. In such a scenario, *if* some policy involving SRM deployment was better than all other available options, then we ought to deploy that SRM technique in that scenario. Despite its moral problems, doing so is morally preferable to all other options. Therefore, in that situation, it ought to be deployed."

This is not a sufficient defense of LAEA, of course, as I have not addressed other important objections to it. Most obviously, we might be skeptical that SRM would indeed be part of the all-things-considered best option in likely pessimistic scenarios. I will address this in the next chapter, providing some explicit reasons for thinking that SRM might be part of the all-things-considered best option in certain pessimistic scenarios. For the sake of argument, however, let us now suppose with the Tentative Climate Engineer that deployment of SRM would indeed be part of the best option available in some pessimistic scenario. Assuming that is true, it seems reasonable to hold that we ought to deploy SRM. A weakness of MDO is that it does not address the Tentative Climate Engineer's reasoning, since the objection merely notes that SRM is morally impermissible even in the kind of case just described. The Tentative Climate Engineer might reply, "Yes, SRM is morally problematic, but what else ought we to do in a pessimistic climate scenario when there is no all-things-considered better option?" The proponent of MDO cannot offer a plausible answer to this question. If some pessimistic climate scenario is a genuine moral dilemma, then there is *no* option that we ought to adopt, since all courses of action would be morally impermissible in that case. This is deeply unsatisfying, and the Tentative Climate Engineer has ground for complaint. One task of moral theory is to provide action-guidance in difficult cases, but MDO seems to undermine any prospect for such action-guidance in those scenarios in which it may be needed most. This is a very serious drawback of viewing pessimistic scenarios as genuine moral dilemmas.

SRM and Agent-Regret

I have examined some reasons to be skeptical of MDO: it is incompatible with a very plausible moral principle, and it undermines moral action-guidance in certain scenarios. However, we should not ignore a great virtue of MDO, namely that it offers an explanation of why both SRM deployment and the figure of the Satisfied Climate Engineer are morally problematic. The proponent of MDO is able to explain these by claiming that SRM deployment would be morally impermissible even if it was the all-things-considered best (or least bad) option available, and the Satisfied Climate Engineer is morally problematic because he acts wrongly in deploying it. This is certainly one way of capturing our intuitions, and if MDO did not have the problems discussed, we might be well-advised to accept it. However, we can explain these phenomena without taking on the disadvantages of MDO. We can do this by maintaining that SRM deployment is an action calling for agent-regret. This allows us to preserve "ought implies can," and it allows for moral action-guidance in pessimistic scenarios.

Associated with Bernard Williams, agent-regret is a type of regret one harbors toward some action of one's own or a collective action in which one has participated. This regret may be informed by thoughts regarding how one might have acted differently than one did in fact act.[17] Agent-regret need not pertain to morally culpable actions. One might feel agent-regret for actions that she is not blameworthy for performing, such as in Williams' example of someone who is driving safely and kills a child who suddenly runs into the road. Agent-regret is therefore not limited to genuinely dilemmatic scenarios. Importantly, agent-regret is not necessarily a phenomenon of wishing that one had acted differently in a given set of circumstances. Instead, agent-regret could involve wishing that those past circumstances themselves had been different, that they had afforded different options, or that the course of action taken had not entailed certain consequences. Williams notes, "Regret necessarily involves a wish that things had been otherwise, for instance that one had not had to act as one did. But it does not necessarily involve the wish, all things taken together, that one had acted otherwise."[18]

In a pessimistic climate scenario, it is at least coherent to judge that some course of action is morally permissible (or even obligatory) and to partake in that action while also harboring agent-regret, since such regret need not involve judging one's action to have been impermissible. The application to SRM is apparent. *If* it was the case in some situation that SRM deployment was part of the best option available, then it would be reasonable to hold that deployment was morally permissible. Yet it would be coherent to regret at the same time that such a drastic measure was needed, as it would be to regret one's own participation in that deployment. One could regret the circumstance of one's participation in the deployment, for example that SRM was needed (as we are supposing for the moment) in the wake of our moral failure to mitigate our emissions. One also could regret the outcome of one's permissible action, for example that SRM deployment likely would result in serious harm and injustice to some parties. Despite these regrets, however, this agent could hold consistently that SRM morally ought to be deployed in those unfortunate circumstances, since agent-regret may be directed toward morally permissible actions.

Agent-regret is not merely an attitude that it is coherent to have in such a case, but it is also an appropriate one. Supposing that SRM is morally preferable to all other options in some case, it is (morally) good to regret both the circumstances that make this the case and the harm and injustice that deployment might cause. It is appropriate to harbor agent-regret here for at least two reasons. First, the context in which SRM deployment is (as we are supposing) morally permissible depends upon serious moral wrongdoing in the past. An agent participating in SRM deployment in such a context should regret that this past wrongdoing has forced her into such a bad situation, even if she is not among those who are guilty of this past wrongdoing, or even if she contributed to the wrongdoing in a miniscule way. Here one wishes that the circumstances had been otherwise. Regretting this fact involves recognizing that SRM likely would be unnecessary in a world in which moral agents had largely complied with their obligations. Agent-regret allows one to countenance the fact that deploying SRM, even if morally justified, signals an enormous moral failure on the part of the many parties who were obligated to reduce their emissions but opted not to do so. This moral failure would then have created a set of circumstances in which it falls to certain agents to adopt a morally problematic option, one that never should have been needed. This surely would be something worth lamenting.

Second, it is plausible that we should harbor agent-regret for those of our actions that create substantial moral disvalue. As I have noted in previous chapters, SRM has the potential to result in serious harm and injustice. One thought in LAEA is that, in a pessimistic scenario, all available options are likely to be bad. Imagine, for example, that all such options likely would involve substantial harm and injustice, but that there is good reason to believe that a policy with SRM would involve less harm and less injustice than any other option. The proponent of LAEA takes it that (assuming all else of moral relevance to be equal) SRM ought to be deployed in that case. But holding this (deontic) view is compatible with holding the (axiological) view that SRM likely would produce substantial moral disvalue. Suppose that SRM is deployed and causes drought in some regions, substantially harming many persons. It is morally bad that SRM results in this harm, even if its deployment was morally permissible. Likewise, it is morally bad for our permissible actions to cause injustice, such as if permissible SRM deployment puts future generations at risk of the termination problem. Even if we grant that LAEA is sound, we nonetheless ought to recognize that SRM could create many instances of moral disvalue. An agent of SRM deployment ought to feel agent-regret for helping to bring about that disvalue. Causing harm or injustice is a bad thing, even if doing so is justified in some case.

The Satisfied Climate Engineer would lack agent-regret, and this is why this figure is morally problematic. Harboring agent-regret for neither the circumstances in which he acts nor the moral disvalue his actions produce, the Satisfied Climate Engineer simply would note that SRM is the best course of action in some pessimistic scenario, conclude that it therefore ought to be deployed, and give no further thought to the moral features of that decision. This figure is rather similar to a driver who, having killed a child through no fault of her own, feels no agent-regret for her action. She might say, "The death is unfortunate, but I regret it in the same way as non-participants of the event regret it—like an onlooker, I view it as a bad thing that the child died. But why should I feel *agent*-regret for the child's death? After all, I was driving at the speed limit, and it was the child's fault for running into the road without looking first." As Williams notes, we would feel "some doubt" regarding a person who reasoned in this way.[19] We tend to think that the driver should regret that *she* killed the child, however unavoidable that death may have been. In conjunction with that regret, she could wish (counter to fact) that she had been able to apply the brakes in time or that the child had not been seriously harmed by the collision. Further, she could recognize the moral disvalue of the harm caused to the child and to those who care about that child. Someone who cavalierly eschews such regret, whether a driver or climate engineer, seems to engage in morally questionable behavior.

I have said that it is morally good to harbor agent-regret even for permissible deployment of SRM, but we might ask the more difficult question of whether doing so is also a duty. It is at least coherent to answer this in the affirmative. We might think, for example, that one has a moral obligation to harbor agent-regret towards any action of his that carries substantial moral disvalue, even if that action is itself permissible. This view is coherent because we can distinguish obligations to act in certain ways from obligations to maintain certain attitudes. On this view, the Satisfied Climate Engineer would be blameworthy for the attitude he holds, even though it is held with respect to an action that (we are supposing) is permissible. The virtue-theoretic approach I offered in chapter four can help us make sense of this. Virtues and vices are tied to attitudes in non-accidental ways, and we can identify some attitudes as inherently virtuous or vicious. Plausibly, the Satisfied Climate Engineer's dismissive attitude toward the harm and injustice done by SRM is itself vicious. We would certainly not expect someone who reliably cares about justice and well-being (i.e., a virtuous person) to maintain such an attitude. Instead, in a pessimistic scenario in which SRM is part of the best option, it is plausible to think that a virtuous climate engineer would maintain an attitude of agent-regret, wishing both that circumstances had been otherwise and that SRM did not produce the moral disvalue that it might. Insofar as the Satisfied Climate Engineer fails to do what a virtuous person would—indeed, doing instead what a vicious person would—we might claim that she violates some duty. This is plausible if we accept that our moral obligations are determined, or at least reliably indicated, by something like the principle, "Do what a virtuous person characteristically would do in the relevant circumstances." This is controversial, however. Fortunately, my claim that an agent-regret account is preferable to MDO does not depend on there being a duty to harbor agent-regret. Instead, I can rely on the weaker, but still sufficient, claim that it is morally good to have agent-regret (and morally bad to lack it) in pessimistic climate scenarios.

The Advantages of an Agent-Regret Account

By relying on this agent-regret account, we can avoid the controversial commitments of MDO without condoning the way in which the Satisfied Climate Engineer goes about deployment. Whereas MDO controversially holds that there are genuine moral dilemmas and undercuts moral action-guidance in those pessimistic climate scenarios in which SRM would be part of the all-things-considered best option, the agent-regret account allows us to disapprove of the Satisfied Climate Engineer while also admitting that SRM may be morally permissible or even obligatory in some contexts. Accepting the latter account opens the following possibility. We might deny that some pessimistic climate scenario would constitute a genuine moral dilemma, hold that there is some permissible action that may be taken in that context, and rely on moral considerations (perhaps drawn from moral theory) to guide our subsequent decision and course of action. We can recognize that a pessimistic scenario is a very bad situation, one created by an enormous moral failure in our past, and we ought to regret that this is the case. Whatever course of action we choose (SRM or otherwise), we can acknowledge that it may result in harmful and unjust outcomes, and we also ought to regret that our action produces such outcomes. However, those regrets need not preclude sufficiently good reasons for acting in certain ways. We can hold that some course of action ought to be taken despite its problems, provided that we feel agent-regret when appropriate.

This approach undercuts much of the motivation for accepting MDO, because it shows that we need not conceive of SRM as morally impermissible in order to account for its problematic features. Gardiner suggests that how we view arguments like LAEA may depend on whether we think there are genuine moral dilemmas. Those who believe in such dilemmas may be "reluctant to consider SRM even as a last resort, and even then are unhappy about having to do so." The concern is that those who endorse LAEA "do not really seem to address the core concerns" of those individuals who believe in genuine moral dilemmas.[20] However, if the foregoing is correct, then perhaps a proponent of LAEA can address these concerns. Unlike the Satisfied Climate Engineer, one might endorse LAEA while feeling the appropriate agent-regret, recognizing the moral problems of SRM while also holding that it ought to be deployed. In this way, we can reject the posture of the Satisfied Climate Engineer without taking on the controversial baggage of MDO. These considerations also suggest that LAEA is defensible in at least *one* respect. If CE really was the part of the all-things-considered best option in some scenario, then it seems reasonable to hold that it would be permissible, perhaps even obligatory, to deploy it in that scenario. MDO implausibly denies this. Conversely, my account is compatible with this particular claim of LAEA. This gives us a reason to prefer the agent-regret account over MDO's position that SRM, like all other options, would be morally wrong in a pessimistic scenario.

Now one might object to the argument of this chapter. First, one might claim that an agent-regret framing misses the fact that pessimistic climate scenarios are morally tragic situations, whereas MDO succeeds in taking this important fact seriously. Whether this objection goes through depends on what is meant by "tragic." As the term is often used in the relevant literature, being faced by a tragic choice in the moral sphere is equivalent to (or at least entails) being faced by a genuine moral dilemma. This is the sense in which Gardiner uses the term. If this is what is meant by "tragic," then it is clear that this objection does not work, because the very point of contention is whether or not a pessimistic climate scenario is best viewed as a genuine moral dilemma. Therefore, I cannot antecedently assume that a pessimistic climate scenario is genuinely tragic, because that would beg the question in favor of MDO. Nor, for the same reason, can I antecedently deny this, because that would beg the question against MDO. Alternatively, perhaps "tragic" is to be used in a different sense, such as to denote a situation that is very bad in some moral respect. But in that sense of the term, it seems that my account does recognize the tragic nature of pessimistic climate scenarios. My heavy reliance on agent-regret is meant to take seriously the fact that SRM deployment would carry substantial moral disvalue, calling on one to regret one's role in such deployment. Even if one disagrees with the agent-regret framing, that framing clearly acknowledges that SRM could be very bad in some moral sense, even if it turns out that SRM is permissible.

Second, one might object that, by supposing pessimistic scenarios in which SRM would be the best option, I have effectively begged the question against MDO in a different way. We might think it obvious that, in a situation in which all options carry substantial moral problems, we ought to take the morally best (or least bad) option. Why assume that SRM would be (part of) the best option in such a case? In response to this objection, it is important to remember that, in the present chapter, I am addressing premise (2) of LAEA: if we are forced to choose among exclusively bad options, then we ought to choose the all-things-considered best of those bad options. I am assuming here that premise (3)—in a likely pessimistic climate scenario, the all-things-considered best option is likely to involve deployment of SRM—is true, but only for the sake of considering premise (2). As I have already noted, chapter six will address the idea behind premise (3). What is interesting about MDO is that it denies premise (2), such that accepting MDO entails rejecting LAEA, even if we suppose that SRM is the all-things-considered best option in some pessimistic scenario. A genuine moral dilemma entails that there is no moral answer to the question of how we ought to act. If so, then it is not the case that we ought to choose the all-things-considered best option. If we find that implausible, then we have a reason to prefer the agent-regret framing I have offered as an alternative to the moral dilemma framing. At any rate, I do not beg the question against MDO by assuming premise (3) of LAEA. A scenario in which SRM is the best (or least bad) of exclusively bad options is precisely the type of case that is relevant to consider here. No one is in favor of deploying SRM when, by one's own lights, better options are available. The question is whether SRM ought to be deployed when it is the best (or least bad) of all available options. An agent-regret framing allows us to answer this question in the affirmative, but without glossing over the (possibly severe) moral problems SRM might carry, such as axiological injustice.

Notes

1. Parts of this chapter are adapted from Toby Svoboda, "Geoengineering, Agent-Regret, and the Lesser of Two Evils Argument," *Environmental Ethics* 37, no. 2 (2015): 207–20.
2. Jason J. Blackstock et al., "Climate Engineering Responses to Climate Emergencies," http://arxiv.org/pdf/0907.5140; David Keith, Edward Parson, and M. Granger Morgan, "Research on Global Sun Block Needed Now," *Nature* 463, no. 7280 (2010): 426–27; David G. Victor et al., "The Geoengineering Option a Last Resort against Global Warming?" *Foreign Affairs* 88, no. 2 (2009): 64–76.
3. Peter J. Irvine, Ryan L. Sriver, and Klaus Keller, "Tension between Reducing Sea-Level Rise and Global Warming through Solar-Radiation Management," *Nature Climate Change* 2, no. 2 (2012): 97–100.
4. Joshua B. Horton, "The Emergency Framing of Solar Geoengineering: Time for a Different Approach," *The Anthropocene Review* 2, no. 2 (2015), 147–151.
5. Stephen M. Gardiner, "A Perfect Moral Storm: Climate Change, Intergenerational Ethics and the Problem of Moral Corruption," *Environmental Values* 15, no. 3 (2006): 397–413.
6. Stephen M. Gardiner, "Is 'Arming the Future' with Geoengineering Really the Lesser Evil? Some Doubts about the Ethics of Intentionally Manipulating the Climate System," in *Climate Ethics*, ed. Stephen M. Gardiner et al. (New York: Oxford University Press, 2010), 284–312.
7. Konrad Ott, "Might Solar Radiation Management Constitute a Dilemma?," in *Engineering the Climate: The Ethics of Solar Radiation Management*, ed. Christopher Preston (Lanham: Lexington Books, 2012), 33–42.
8. Gardiner, "Is 'Arming the Future' with Geoengineering Really the Lesser Evil?," 301.
9. We also could reject both, of course, but I will not examine a case for doing so here.
10. Toby Svoboda, "Is Aerosol Geoengineering Ethically Preferable to Other Climate Change Strategies?" *Ethics & the Environment* 17, no. 2 (2012): 111–35.
11. Michael J. Zimmerman, *The Concept of Moral Obligation* (New York: Cambridge University Press, 1996).
12. Terrance C. McConnell, "Moral Dilemmas and Consistency in Ethics," *Canadian Journal of Philosophy* 8, no. 2 (1978): 269–87.
13. Christian Baatz, "Responsibility for the Past? Some Thoughts on Compensating Those Vulnerable to Climate Change in Developing Countries," *Ethics, Policy & Environment* 16, no. 1 (2013): 94–110.
14. Although cf. the Arctic Methane Emergency Group, a group of scientists calling for serious consideration of deploying SRM in the near future: Arctic Methane Emergency Group, "AMEG's Declaration," http://ameg.me/index.php.

15. Keith, Parson, and Morgan, "Research on Global Sun Block Needed Now"; Blackstock et al., "Climate Engineering Responses to Climate Emergencies."
16. Adam Millard-Ball, "The Tuvalu Syndrome," *Climatic Change* 110, nos. 3–4 (2012): 1047–66.
17. Bernard Williams, *Ethics and the Limits of Philosophy* (Cambridge: Harvard University Press, 1985), 123.
18. Ibid., 127.
19. Ibid., 124.
20. Gardiner, "Is 'Arming the Future' with Geoengineering Really the Lesser Evil?" 302.

6 Comparisons[1]

We have seen that SRM is ethically problematic in some deeply disturbing ways. But given that we should consider climate justice through the lens of non-ideal theory, and given that we should not view cases in which SRM is potentially attractive as genuine moral dilemmas, there may be situations in which deployment of some SRM policy is ethically permissible despite these problems. This is plausible if our approach to ethical assessment of climate policies is comparative. On such an approach, roughly, an ethically problematic climate policy can be permissible if it compares favorably to other available options. In a case in which no available policy avoids having serious ethical problems (a pessimistic scenario), it is not enough simply to point out some policy's problems and conclude that it is therefore prohibited, for this would require us to deem all the available policies to be prohibited, yielding a genuine moral dilemma. The question of what we ought to do in a pessimistic scenario is a pressing one, calling for moral action-guidance, something that a moral dilemma framing would foreclose. Unfortunately, a pessimistic scenario is precisely what we can expect in the future, unless we very soon pursue drastic and rapid emissions mitigation. Without such an abrupt change in our global climate policies, at some point (perhaps one already reached) past emissions will have committed us to very dangerous climate change. In a scenario of that type, all courses of action would plausibly carry substantial ethical problems: emissions mitigation cannot address committed climate change; adaptation is limited in the climatic harms it can ameliorate, and even successful adaptation may impose heavy burdens of its own, including directing resources that otherwise might be used to meet basic needs; and, most obviously, continuing with business-as-usual would simply permit the harm and injustice of climate change to accumulate unchecked. In short, although there are serious concerns about SRM regarding risks of harm, distributive injustice, and procedural justice, the alternatives options in the future are likely to carry serious problems as well. In that case, I will argue, we have good ethical reason to compare the available policies and choose that option which is best (or least bad, as the case may be). Relying on this, I will then argue that certain uses of SRM are plausibly permissible in some pessimistic scenarios that could reasonably hold in the future.

Acting in Pessimistic Scenarios

I mentioned that, in pessimistic climate scenarios, we ought to adopt that course of action which is best (or least bad). By this I mean the course of action that is *morally* best, all things considered. Accordingly, we cannot determine the best option simply by weighing up the aggregate harms and benefits of each policy and comparing the net results. As we saw in chapter one, there are daunting epistemic problems when it comes to performing this accounting work, and this approach would overlook matters of great moral importance, namely those of justice. In order to determine the best (or least bad) course of action, we need to weigh up all the relevant moral goods and ills of each available course of action, then compare the results across available courses of action. Of course, this is not a matter for simple calculation. Because it involves taking into account features that are difficult to compare (and arguably incommensurable), this exercise unavoidably involves judgment. I do not suggest that there is some decision procedure (in the algorithmic sense) for performing this task. This opens the possibility that even reasonable, well-informed persons will reach divergent verdicts on which available policy is actually the best (or least bad) in some pessimistic scenario. The most straightforward way to view this is as an epistemic problem. Although there presumably would be some fact of the matter regarding which policy is actually the best, we might not be in a position to know this fact. Of course, in some cases, the available option that is morally best might be obvious, or at least sufficiently clear on reflection. Given the possibility of such cases, we might be able to avoid the aforementioned epistemic problem in some pessimistic climate scenario that happens to hold in the future. It would be naïve to assume that this fortunate outcome will hold, however, so we must address the epistemic problem here. For dealing with difficult cases like this, I suggest two options: relying on virtuous agents and deferring to procedurally fair decisions.

The likelihood of pessimistic scenarios in the future gives us yet another reason for wanting decision-makers to be virtuous. As we saw in chapter four, a virtuous agent is someone who is sensitive to morally salient features, reliably motivated to act in morally valuable ways, and (via practical wisdom) good at discerning the best ways of so acting. A non-virtuous agent will have difficulty in navigating a pessimistic scenario when the morally best course of action is not obvious. He may be insufficiently sensitive to some morally relevant feature, or he might have little idea of how to incorporate competing considerations into a coherent course of action. Now if deciding how to proceed was simply a matter of applying some decision procedure, we would not need our decision-makers to be virtuous. In that case, we would require only technical competence in running the procedure properly. But determining appropriate climate policy in a pessimistic scenario is not a matter for decision procedures, especially if we take seriously the various moral considerations we have noted so far. This could be so for two distinct reasons. First, as a matter of principle, it may be that complex moral decisions are not amenable to decision-procedural resolutions. Second, it may be that decision-procedural resolutions in pessimistic scenarios are possible in principle, but that they are impracticable, either because we lack sufficient information to feed into some procedure or because the procedures available to us are insufficiently sophisticated for the task at hand. On the former view, there simply is no *possible* algorithm that will tell us how to act in cases of the relevant kind. On the latter view, while there is some possible algorithm that could perform this work, we will be unable to utilize it in practice—for example, because there is too much uncertainty regarding the appropriate inputs. I do not take a position on whether either of these views is correct. Whatever the reason, it is difficult to deny that decision procedures will not help us determine what course of action to take in pessimistic climate scenarios. Anyone attempting to think through what to do in such a situation must grapple with the many moral features that are in play. This is precisely what the practically wise, virtuous agent is skilled at doing. Of course, there is no guarantee that the practically wise, virtuous agent will succeed in determining which available policy is best. Such agents are not infallible, so the possibility of mistaking a sub-optimal policy for the optimal one is unavoidable. Nonetheless, we are better off relying on virtuous agents and their practical wisdom than relying on non-virtuous (or vicious) agents. All else being equal, the former type of agent has much greater prospect of success.

A second way of dealing with the epistemic problem—namely, of determining what course of action is morally best (or least bad) in pessimistic scenarios—is to let just procedures determine what we are to do. This relies on the pragmatic reason I offered in chapter three for favoring procedurally just policy-making. When there is uncertainty and reasonable disagreement about which policy is, all things considered, the best one in a moral sense, it can be useful to let those with legitimate claims on decision-making to work out collectively how to proceed, with appropriate parties contributing to the ultimate decision in proportion to the strength of their respective claims to do so. This provides a principled way to navigate disagreement and provide divergent views fair hearings. It is not unreasonable to hope that such a procedurally fair process might rule out extreme and idiosyncratic proposals. For instance, because of the variety of perspectives included in fair decision-making, some parties might have the opportunity to point out merits or deficiencies that other parties might otherwise have overlooked, such as risks of harm that might affect some regions but not others. As with virtuous agents, it cannot be guaranteed that the result of this process will succeed in identifying the option that is indeed the best one. Nonetheless, for the reasons given, it can be helpful to defer to procedurally just decision-making in order to deal with uncertainty on this matter.

Admittedly, there appears to be a potential tension between these two ways of dealing with uncertainty about the best available policy in a pessimistic scenario. It is possible that the judgment of some practically wise, virtuous agent will diverge from some collective judgment reached in a procedurally just manner. While this is possible, we should notice that procedurally just decision-making can include input from virtuous agents—constitutive input if virtuous agents themselves have legitimate claims, and advisory input if those agents lack such claims—as well as that parties with legitimate claims can choose to rely on this input. Because they themselves are just, virtuous agents will respect the processes of procedural justice. In cases of decision-making in which just persons are somehow involved, this gives us reason to think that (non-ideal) procedural justice will be respected to some extent. Nonetheless, there can be divergence between the judgment of virtuous agents and the judgment of those with legitimate claims, which may indicate that we have a particularly difficult case. I have suggested only that relying on virtuous decision-makers and deferring to procedurally just decisions are potentially useful tools in deciding what policy is best (or least bad) in a pessimistic scenario, and there is no assurance that the results of utilizing these tools will always be in harmony. That, of course, does not mean the tools themselves are useless.

Before continuing, I should acknowledge another potential objection to my approach, namely that it is too demanding to require that the *best* (or *least* bad) option be chosen in pessimistic scenarios. After all, in many cases it is obvious that sub-optimal choices are permissible. For instance, although it would be morally better for us to spend all of our spare time furthering worthy charitable causes, on most accounts we are permitted to spend some of that time relaxing instead. This, of course, is to draw a distinction between the obligatory and the supererogatory. With that in mind, we might ask: why ought we to prefer the best policy rather than one that is merely sufficiently good? In answer to this objection, the key is to recall that we are dealing with pessimistic scenarios, or situations in which all available courses of action have serious moral problems. Hence, to choose the best option in a pessimistic scenario is to choose an option that nonetheless has serious moral problems. This is not plausibly described as supererogatory. Rather, it is plausible to think that we are obligated to choose the best (or, perhaps more appropriately, least bad) option in those cases in which each alternative carries serious (and, compared to the best option, worse) moral ills. In non-pessimistic scenarios, the best option might often be very demanding, and sub-optimal options might carry no moral problems at all. Plausibly, we are permitted to favor some sub-optimal option under such conditions, but only if they lack serious moral problems. By definition, this is not the case in pessimistic scenarios. Hence, with one caveat to be mentioned, it is not implausible to require that the best option be taken in such scenarios.

The caveat in question is related to the fact that, for the reasons I have noted in previous chapters, pessimistic climate scenarios are likely to fall within the realm of non-ideal theory. That being the case, decision-makers within a pessimistic scenario would be limited to those options of climate policy that are both politically feasible and likely to be effective. Accordingly, my view is that, in a situation in which all available courses of action carry substantial moral problems, we ought to choose the best (or least bad) option that satisfies conditions of both feasibility and effectiveness. As we have seen, there might be some conceivable policy that is morally preferable to all others but which is either unlikely to work or is a political non-starter. In line with non-ideal theory, when I refer to the "best" policy in a pessimistic scenario, I mean the policy that is best (from the point of view of justice) within the domain of policies that are feasible and likely to be effective. In determining which policy within this class is the best one, we turn to the moral permissibility condition of non-ideal theory. Clearly, our central question here is a comparative one. For this reason, it is the comparative component of the permissibility condition—rather than the proportionality component—that is of central importance in pessimistic scenarios.

Because of this, the proportionality criterion, although still relevant, plays a different role here than in ordinary cases. Usually, some policy will count as non-ideally just (in the deontic sense of being permissible) *only if* the moral good it secures is proportional to the moral ill it imposes. In pessimistic scenarios, such proportionality may be impossible to secure, at least with actions that are politically feasible and likely to be effective. Suppose that in some situation all available options—including inaction—carry such substantial moral ills that there is no feasible or effective way to secure proportionality. In such a case, the best (or least bad) option would not satisfy the proportionality criterion. However, as I argued in the previous chapter, it is problematic to view this best option as being morally impermissible, because doing so runs counter to very plausible moral principles (e.g., "ought implies can") and undermines moral action-guidance precisely when we need it most. In a bad situation like the one just mentioned, it is more plausible to treat the best option as morally permissible (in a non-ideal sense) despite the fact that this option fails to secure proportionality. As I also argued in chapter five, adopting this course of action calls for agent-regret on account of the moral ills it imposes, but doing so is permissible because it is better than any other (feasible and effective) option. Accordingly, if some pessimistic scenario precludes proportional responses, we should not treat proportionality as a necessary condition for some response to be non-ideally just in the deontic sense. Instead, we should consider how (feasible and effective) policies compare to one another in terms of the moral goods and ills they carry. This does not mean that proportionality is to be ignored in such pessimistic scenarios, because considerations of proportionality are still relevant for purposes of comparing policies. Among several non-proportional responses, the respective ratios of goods to ills might vary substantially. That is, although some policy response might entail moral ills that are disproportionate to the moral goods it secures, the alternatives might be disproportionate to an even greater degree. In that case, the first disproportionate policy would compare favorably to the others.

Climate Policy and Non-Ideal Theory

Despite its own ethical problems, there are reasons to think that some policy involving SRM could compare favorably to other options in certain pessimistic scenarios. Because SRM has the potential to cool the climate quickly, it might be able to manage some climate risks better than slower-acting responses, such as those relying heavily on mitigation and adaptation. This is a major advantage when it comes to securing beneficial outcomes while minimizing harmful ones, as well as easing potentially unjust distributions thereof. Importantly, I will not defend the view that an "SRM-only" policy would be attractive in pessimistic scenarios. While virtually no one would defend it anyway, it is very unlikely that such a policy would be the best (in a moral sense) one available in any (reasonably likely) pessimistic scenario. This is because an SRM-only response would not address the problem of future emissions and the risks of harm and injustice that such emissions would carry. All else being equal, a policy that reduces these future risks is better than one that leaves them unaddressed. Instead, I will sketch a type of hybrid policy, which involves short-term SRM in order to provide time for emissions mitigation and adaptation—and, possibly, affordable CDR at a large scale. This has advantages over an SRM-only policy. To take one example, it involves less intergenerational injustice, because it does not put future generations indefinitely at risk of abrupt termination of SRM.

Under ideal conditions in which all moral agents fully complied with their duties of justice—or at least came fairly close to doing so—we would not be justified in deploying SRM. Given the risks of axiological injustice associated with SRM, and given the availability of apparently better (from the point of view of justice) options—such as those involving some combination of mitigation, adaptation, and perhaps financial assistance to less developed countries to transition to renewable sources of energy—it is reasonable to think that SRM would not be permissible as a matter of ideal justice. When possible, we have ethical reason to prefer an approach to climate policy that is likely to secure morally appropriate distributions of burdens and benefits and to render proportional satisfaction of all legitimate claims to participation in decision-making on relevant matters. But SRM is receiving serious attention precisely because these ideally just responses to climate change are not being pursued, at least not to a sufficient degree. High emissions on the part of some human beings in the past, present, and likely future will commit us to substantial warming, posing risks of substantial injustice and harm to present and future generations. Despite some apparent progress (e.g., in the Paris Agreement) in securing national commitments to mitigation targets, the world is nonetheless on track for atmospheric greenhouse gas concentrations that are likely to be very dangerous in the future. This is because many emitters are doing far too little to mitigate their emissions. This constitutes a substantial moral failure on the part of many of us. More specifically, many emitters are failing to comply with their duties of justice—as well as failing in their duties of beneficence—to present and future generations. Under such non-ideal conditions, SRM may do better than emissions mitigation or adaptation alone when it comes to securing morally valuable aims, such reducing distributive injustices and harms that climate change might otherwise cause. For instance, at some point in the future, SRM could be used to curb and to slow the dangerous warming to which past emissions will have committed us.[2] Depending on its severity, such warming might damage what I call "climate-sensitive goods," thus affecting both the welfare of relevant parties and the ways in which those goods are distributed. By a "climate-sensitive good," I mean any good that is susceptible to being affected (e.g., by increasing or decreasing its quantity or quality) by climate change, whether that climate change is facilitated by anthropogenic emissions, CE technologies, or natural variability.

There are limits to what mitigation and adaptation can do to prevent the erosion and unjust distribution of such goods. Due to committed warming from past emissions and inertia in the climate system, we may face a future scenario in which all available policies would result in substantial net harm.[3] In such a scenario, this harm could outstrip the adaptive capacities of human communities—even if, in principle, a society's adaptive limits are contingent[4]—and mitigation of future emissions would not help prevent committed warming from past emissions. Conversely, SRM could have a fast impact on the climate system, potentially curbing warming within several years.[5] In a bad future scenario of this sort, there is a *prima facie* plausible case that SRM could reduce net harm relative to a response including only adaptation and mitigation. Because climate harms tend to impact those who are worst off (and least responsible for climate change) disproportionately, reducing the net harm of climate change might also ease climate injustice. So far, these suggestions are merely speculative.

As we saw in chapter two, a major ethical concern about SRM is that it carries substantial risks of injustice. It is important, however, to consider various specific types of SRM deployment, some of which might carry reduced risks of injustice relative to other types. It will not do to point out ethical worries about SRM in some of its guises and then reject SRM as such on that basis, as policies involving SRM can differ in many significant ways. For instance, it would be reasonable to be very skeptical of a policy that would deploy SRM indefinitely or treat it as a strict alternative to mitigation. Due to the lack of an exit strategy, a policy like this could entail a long-term commitment to the harms and injustices that happen to accompany SRM.[6] Fortunately, no serious commenter favors an SRM-only policy. That being the case, we risk critiquing a straw man if we argue against such a policy. At best, thinking of SRM in isolation from other components of a hybrid climate policy is useful as a kind of thought experiment, highlighting ethically problematic features that we should want to avoid in any actual SRM policy proposal. It is more interesting, however, to think about policies that are designed to be sensitive to the risks of injustice that SRM might carry. If these risks can be successfully ameliorated, then a major ethical concern regarding SRM will lose some of its force. Presumably, no SRM policy would ever be fully unproblematic in an ethical sense—but neither would mitigation or adaptation if past warming had committed us to dangerous (e.g., harmful and unjust) warming. This is especially so if, as we ought, we take into account the potential for some mitigation policies to slow poverty eradication that might otherwise have been feasible through reliance on fossil fuels. In some cases, assuming *relatively* low risks of injustice, SRM's ability to cool the planet rapidly might tip the balance of justice considerations in its favor over policies that lack SRM.

Importantly, although my account will depend upon the idea of pessimistic scenarios, this line of reasoning does not require the controversial appeal to the idea of a climate emergency. This latter idea often relies on the possibility of abrupt changes in the climate (e.g., tipping points) that precipitate very harmful outcomes.[7] As we have seen, the once-common appeal to climate emergencies has come under scrutiny, both because of epistemic concerns (e.g., that we are unlikely to be in a position to have justified beliefs on whether some tipping point is imminent) and political concerns (e.g., regarding the potential abuse of emergency appeals for political reasons).[8] These are very reasonable concerns. Fortunately, the notion of a pessimistic climate scenario need not involve appeals to emergencies in general, nor to abrupt climatic changes in particular. Instead, it may be the case that future warming promises to erode climate-sensitive goods in a linear, gradual fashion. All else being equal, the erosion of that good would be a bad thing. Depending on the situation, we may have ethical reasons (e.g., grounded in duties to promote justice, duties to promote welfare, or both) to halt, slow, or reverse such erosion. Unfortunately, there can be cases in which every way of doing so—at least, every way that is politically feasible and likely to be effective—is morally problematic in some substantial way. This is what I call a pessimistic scenario. What makes the scenario pessimistic is the fact that there is no (feasible, effective) way of averting substantial moral problems. On the question of whether some scenario is in fact pessimistic in this sense, it matters not whether the moral problems owe to abrupt or gradual changes. So appealing to pessimistic climate scenarios is distinct from appealing to climate emergencies.

How Might Risks of Climate Injustice Be Reduced?

Throughout this book, I have acknowledged deep ethical problems with SRM, including risks of substantial injustice, both distributive and procedural. However, we should also acknowledge that these problems may be present to varying degrees—and, in some cases, not at all—depending on what variety of SRM is used, the context in which it is deployed, and (crucially) what other policy components accompany it. Importantly, SRM policies could pose substantially different degrees of risks of injustice. We should look for ways of reducing these risks. There are several reasons for doing so. First, doing so goes hand in hand with offering an intellectually honest assessment of SRM. It would be easy to dismiss a straw man SRM policy—say, an SRM-only policy that lacks even any aspiration to mitigation or adaptation. Obviously, in addition to being a rather uninteresting venture, dismissing *that* policy would be unfair to those who think we should take SRM seriously. At least at the current time, virtually no proponent of SRM research favors a climate policy like that just described. Instead, we should consider the merits and deficiencies of SRM policies that stand to have actual supporters, such as those that are responsive to the risks of injustice that are involved. Second, given serious interest from prestigious research institutes and the IPCC,[9] as well as calls for near-term deployment,[10] there is a reasonable prospect that SRM will be deployed in the future. Whether or not we judge that to be a good idea, it is wise to think about ways of reducing the ethical problems of SRM, for the simple reason that—setting aside the question of permissibility—a less problematic policy is to be preferred to a more problematic one, all else being equal. Third, and most directly for the purposes of this chapter, conceiving of SRM policies with reduced risks of injustice is central to the task of determining whether SRM might be part of the best policy option in a pessimistic climate scenario. Recalling that I am arguing that, in such a scenario, we are obligated to adopt the best (in a moral sense) option that is both politically feasible and likely to be effective, and we must address the question of whether there is some (feasible, effective) version of SRM that does better than other versions on this front. If we decline to consider that question in a serious manner, we run a risk of failing to identify the (feasible, effective) policy option that would be best. For these three reasons, and the third in particular, we should look for ways to reduce risks of injustice associated with SRM.

First, we might ask if SRM deployment could be technically designed such that it substantially reduces risks to particularly vulnerable parties. Computer model simulations suggest that SRM's impact on precipitation patterns could depend on both the location and quantity of aerosol injected into the stratosphere. For instance, Haywood et al. find that "large asymmetric stratospheric aerosol loadings concentrated in the Northern Hemisphere are a harbinger of Sahelian drought whereas those concentrated in the Southern Hemisphere induce a greening of the Sahel."[11] This indicates that, compared to Northern Hemisphere deployment of SRM, deployment in the Southern Hemisphere could involve a reduced risk to parties living in the Sahel. Further, Irvine et al. find that SRM's impact on the hydrological cycle could be tied to the quantity of aerosols injected, with lower quantities leading to reduced impacts on precipitation.[12] If that is the case, then less aggressive cooling meant only to slow (rather than reverse) warming may be less risky than more aggressive cooling.[13] This suggests that, depending on their technical design, some SRM policies could carry a reduced risk of some injustice to some parties, whether by reducing the overall risk of harm to all parties or by shifting risks to some parties (e.g., high emitters who can afford to adapt to precipitation change) rather than others (e.g., low emitters who cannot afford to adapt and thus are particularly vulnerable to the hazard in question). For example, from the perspective of distributive justice, we could have an ethical reason to prefer Southern Hemisphere deployment to Northern Hemisphere deployment, as the former may pose less risk of (potentially unjust) harm to vulnerable parties through the erosion of certain goods that are sensitive to alterations in precipitation patterns. Of course, the foregoing are just possible examples. Further research would be needed to determine what other risks of injustice Southern Hemisphere deployment might carry (including unaddressed risks of injustice due to anthropogenic emissions), as well as how such risks compare to those of other SRM policies. The general point, however, is that technical design decisions might go some distance in ameliorating ethically problematic features of SRM. Given a set of possible SRM policies, we might investigate their respective risks of injustice, attempting to identify that (feasible, effective) SRM policy with the lowest such risk, as well as identifying further areas of research that might prove helpful in determining this.

Second, as Wigley has proposed, SRM might be coupled with emissions mitigation, with the former serving as a short-term response meant to "buy time" for instituting sufficient cuts in emissions, after which SRM could be phased out.[14] Such a policy potentially could reduce risks of injustice to future generations, since it would not require SRM to be maintained into the distant future. Compared to a stand-alone SRM policy, this hybrid policy would decrease the number of future persons put at risk due to the termination problem. While such an SRM policy still could put near-term future parties at risk of unjust harm, this presumably would be ethically preferable to a stand-alone SRM policy that would put a far greater number of future parties at such risk. Of course, there is a wide range of possible hybrid policies that include SRM. In addition to mitigation, such a policy might also include adaptation and CDR. For instance, we can imagine a policy that utilizes "low-grade" stratospheric aerosol injections to slow temperature increase while carrying a relatively low risk of unjust precipitation change, commits various parties to mitigation efforts that gradually ramp up over time, finances adaptation in less developed countries, and utilizes some variety of CDR (e.g., BECCS) as a wedge in the policy meant to reduce the risks arising from past emissions. Obviously, for any such policy there will be questions of political feasibility, particularly when it comes to paying the costs of CDR and adaptation, and the answers to such questions will depend in part on the details of the particularly policy on offer. But there is reason to think that a hybrid policy of this kind could be both effective at managing emissions-related injustice and morally permissible. At the very least, when it comes to effectiveness and permissibility, such a hybrid policy has clear advantages over an SRM-only policy, because the former can address morally relevant matters that the latter cannot address (e.g., managing the risks of ocean acidification). Now it might turn out that the best (in an all-things-considered moral sense) hybrid policies are not politically feasible. Indeed, this is to be expected. For instance, some powerful parties might be unwilling to pay for large-scale CDR or to finance ambitious adaptation in less developed (and more vulnerable) countries. Still, powerful parties might be willing to pay for *some* degree of CDR or adaptation in less developed countries, and this might make feasible certain hybrid policies that are, from a moral point of view, preferable to non-hybrid policies that happen to be politically feasible.

While certain technical design decisions could reduce risks of injustice associated with SRM, we may safely assume that these measures would be insufficient to render SRM deployment perfectly just in the axiological sense. Accordingly, we should ask how SRM-related injustice should be addressed and (if possible) rectified. As we have seen, a compensation regime could play an ameliorative role here. I made the case in chapter two for an SRM compensation system in which polluters (understood to be agents of SRM who are also high emitters of greenhouse gases) remunerate parties that (unjustly) fall below an impersonal baseline of well-being due to conditions or events caused by SRM, with the amount paid being equal to that needed to return such parties to the impersonal baseline. I argued that this type of SRM compensation regime does well by the light of non-ideal theory, meaning that it is politically feasible, likely to be effective, and morally permissible. If so, then a compensation system of this kind could go some distance in ameliorating the injustice of SRM policies, including hybrid variants thereof. Accordingly, we should take this into account when evaluating whether some potential policy involving SRM is (morally speaking) the best available policy that is both effective and politically feasible. We should not forget the caveats I noted in chapter two, however. It is virtually certain that SRM compensation would be imperfect—for example, because some harms (e.g., cultural losses) are not plausibly taken to be susceptible to economic remuneration. Nonetheless, all else being equal, an SRM policy (hybrid or otherwise) that includes compensation for victims of injustice would be ethically preferable to an SRM policy that provided no such compensation. In a pessimistic scenario, it is possible that there would be some sophisticated hybrid policy involving SRM that is politically feasible, likely to be effective, and (thanks in part to compensation) morally favorable to other options.

In addition to risk-reducing technical design decisions, we also should look for ways to reduce the risk of procedurally unjust decisions regarding SRM deployment. We might think of such measures as attempts to increase compliance with duties of procedural justice. As I noted in chapter three, the international community might adopt (and enforce) a treaty prohibiting unilateral (or otherwise exclusionary) decisions to deploy SRM. I also noted Horton's suggestion that techniques (e.g., persuasion) from "international management theory" might be used to foster multilateral decision-making with regard to SRM technologies, including the question of whether to deploy SRM in some case.[15] Depending on how such techniques are used, they might help to foster a greater degree of incomplete fairness than might otherwise prevail. For instance, in contrast to imposing economic sanctions or threatening the use of force, persuading parties to adopt some climate policy is at least compatible with honoring their claims on decision-making power, as parties who seek to persuade one another can allow for at least partial satisfaction of their respective claims. In line with what I argued in chapter three, this opens the possibility for non-ideal procedural justice when it comes to SRM. Caney suggests several other ways in which we might increase compliance with duties of climate justice in general, some of which are potentially relevant to procedural justice with regard to SRM.[16] In addition to sanctions for non-compliance, he mentions fostering appropriate norms and engaging in civil disobedience. The idea is that such practices can put pressure on parties not to behave in (procedurally) unjust ways. For instance, if the international community develops and maintains a strong norm against unilateral (or small-scale multilateral) decision-making on SRM, it will be somewhat more difficult for parties to engage in in such decision-making. Violating such norms might carry unwelcome implications for the transgressing parties, such as displeasing allies and potential trade partners. Likewise, through protest or forms of civil disobedience, populations might move their governments toward great compliance with requirements of procedural justice, such as by demanding that officials abide by the promises made as part of the UNFCCC.

Now in one sense, these measures might not increase compliance with duties of procedural justice strictly understood. If one abstains from some unjust action solely due to social pressure brought to bear on him by others, we might think that—to borrow from Kant—he is acting merely in accordance with justice rather than acting from the motive of justice. Depending on one's commitments in ethical theory, this might preclude that person's choice from being truly just. There are two responses to this. First, it is possible that social pressure could help some parties who previously overlooked procedural justice to become aware of its relevance and importance in some case. Displays of civil disobedience, for example, might highlight the presence of some deep unfairness that one had previously not perceived. At least to some extent, the social pressure exerted by such a public display might tap into one's moral motivation, persuading her to act from the motive of justice going forward. Second, even if it is the case that some parties do not care about justice very much—or, in extreme cases, at all—we still have good moral reason to want to increase the prevalence of practices on their part that merely accord with justice. After all, the crucial task for clinical-theoretic approaches to non-ideal justice is to avert and mitigate impending injustices. If we can do this by using (permissible forms of) social pressure to persuade parties to avoid or reduce unjust practices, then it is desirable to do so, even if the persuaded parties lack any internal motivation to care about justice. Taking a Kantian line, it may be that the pressured parties are not really behaving in a just fashion when they succumb to such techniques, but none-theless their doing so averts injustice that otherwise might have occurred, namely a failure to accord proportional decision-making to parties that have legitimate claims on decision-making power.

Might an SRM Policy Satisfy the Conditions of Non-Ideal Justice?

Some uses of SRM are plausibly taken to be politically feasible, largely in virtue of their susceptibility to small-scale, multilateral deployment, as well as their potential to carry relatively low direct costs. As we have seen, unlike large-scale emissions mitigation, successful deployment of SRM requires neither broad agreement nor coordination among the various parties who have in interest in the matter. Among other things, it is the need for such broad agreement and coordination that has made it difficult to achieve the substantial mitigation needed to avert risks of severe climate injustice, the modest success of the Paris Agreement notwithstanding. A major impedi-ment to reaching a binding agreement on substantial mitigation is the (short-term) costs of transitioning away from fossil fuels to renewable forms of energy. So far, some politically important, high-emitting parties have been unwilling to pay these costs, and this has rendered aggressive mitigation politically infeasible. While some parties might be willing to pay the costs involved, the required degree of mitigation cannot be achieved without the cooperation of high emitters. In principle, some types of SRM can avoid these problems. As we have seen, CE via stratospheric aerosol injections is projected to be cheap in comparison to aggressive mitigation, and a small group of countries could effectively deploy SRM without securing the con-sent or cooperation of others.

These observations do not provide sufficient justification for my position, however. My claim is that, in some reasonably likely pessimistic scenarios, there will be some *hybrid* SRM policy that satisfies the requirements of non-ideal justice. From the mere fact that SRM is politically feasible in some case, it obviously does not follow that a hybrid policy involving SRM is politically feasible in that case. This being the case, one might grant that SRM is polit-ically feasible on its own, yet contend that a hybrid policy involving both SRM and mitigation is not politically feasible, given the latter component. However, there is reason to think that, on a hybrid policy including SRM, a commitment to mitigation will be more feasible than it would be absent SRM. If the type of SRM used is successful at reducing risks by slowing the rate of climate change, then this policy can buy time for the mitiga-tion it requires. For instance, we might begin with relatively modest cuts in emissions that gradually ramp up over time, with SRM moderating the warming (and attendant risks) that would otherwise follow from those emis-sions, as well as warming from historical emissions. While many parties might be unwilling to pay the costs of ambitiously rapid mitigation, they might be willing to pay the (presumably greatly reduced) costs of far less ambitious, gradual mitigation. In fact, there are encouraging indications that society may already be transitioning to renewable forms of energy, so SRM might buy time for such mitigation to occur while reducing risks of climatic injustice.[17] A similar possibility holds for other aspects of a hybrid policy, including adaptation and CDR. Adequate adaptation to climate change might be politically infeasible in a rapidly warming world, given the high costs of doing so at the pace required, but adequate adaptation might be feasible in a world that warms more gradually. This is to say that SRM might buy the time needed in order to make adequate adaptation feasible. Likewise, SRM might buy the time needed for some types of CDR to become feasible thanks to falling costs over time. To be sure, these are merely possibilities. It would be foolish to predict, for example, that the costs of some type of CDR will in fact drop to a level that makes it feasible within some relevant timeframe. This, of course, would depend on various uncertainties, including the point in time at which some pessimistic scenario arises. Nonetheless, considerations of this kind show that it is plausible to think that successful SRM would enhance the prospect of mitigation, adaptation, or CDR proving politically feasible. If for no other reason, it is plausible to think this because of SRM's potential to manage climatic risks and buy time for other components of a hybrid policy to play their parts.

Now one might object to this line of reasoning by claiming that successful SRM, simply in virtue of being successful, would undermine other components of a hybrid climate policy. This would be to adopt some version of a risk compensation (sometimes called "moral hazard") argument against SRM deployment.[18] In general, such arguments express the concern that, if SRM succeeds (or seems to succeed) in reducing climatic risks, some parties will compensate for this (perceived) reduction in risk by taking on some degree of additional risk. In the present context, the concern would be that SRM's reduction of climatic risk would result in some parties not pursuing other measures of any hybrid policy. For example, some parties might be less motivated to mitigate their emissions if SRM seems to protect them from the ill effects of emissions. Depending on how strong and widespread such risk compensation would be, it might undermine the political feasibility of some hybrid policy, given that one component thereof could undercut other components, preventing the enactment of that policy as a whole. However, although risk compensation is a frequently mentioned concern in the literature on SRM, there is little evidence that it would be operative in the case of SRM. As I noted in chapter two, there have been some studies conducted that survey individuals' expected responses to SRM, but we cannot use these to infer much about risk compensation with respect to SRM hybrid policies. This is not only because individuals might be unreliable when it comes to predicting their own future behavior, but also because the motivation and behavior of individuals might not be a reliable indicator of commitment to mitigation at the societal level. Of course, it is possible that future evidence will come to light showing that successful SRM likely would entail risk compensation of a sort that would undermine the political feasibility of hybrid policies. However, this would only be a problem for my view if the risk compensation effect proved to be substantial. In some cases that have been studied, the benefits of risk-reducing technologies (e.g., seatbelts) outweigh the additional risks brought on by new behavior (e.g., speeding).[19] Further, it may be that one's willingness to tolerate additional risk is rather limited. With a seatbelt, someone might drive five or ten miles per hour faster than she would without a seatbelt, but most are unlikely to drive forty or fifty miles per hour faster with a seatbelt. If something similar holds for successful SRM, then the risk compensation that follows deployment is unlikely to render hybrid policies infeasible. For instance, assuming the deployment of risk-reducing SRM, we might be *somewhat* less willing to pursue mitigation, but we might remain *sufficiently* willing to do so, such that mitigation would be politically feasible on the timescale required by the hybrid policy in question. Once again, it is possible that risk compensation associated with SRM would be especially strong, effectively rendering hybrid SRM-mitigation policies politically feasible. However, given our current knowledge, it is reasonable to think that some hybrid policies involving SRM would be politically feasible.

Further, certain types of SRM are likely to be effective in achieving their aims, thanks to their capacity to cool the planet, as indicated by modeling studies and observations of natural analogues (e.g., volcanic eruptions). To be sure, there is ample uncertainty regarding the impacts of SRM, especially at the regional level, regarding such important matters as changes to the hydrologic cycle and ozone depletion. Without field tests, which have yet to be conducted, there will be severe limits to how confident one should be about SRM's specific impacts. Nonetheless, it is hard to deny that SRM appears capable of inducing global cooling, and it is hard to deny that such cooling would help reduce some of the risks tied to a warmer world, all else being equal. Indeed, it is precisely this potential that moves some scientists to call for large-scale field testing of SRM.[20] Of course, future field tests or other research might reveal serious problems, including currently unknown ones, with SRM technologies. Depending on their severity, these problems might call SRM's likely effectiveness into question. This is also sometimes cited as a further reason to conduct field tests: if some SRM technology will not work as some hope, it is best to discover that fact soon, so that we may turn to more promising options for addressing climate change.[21] Nonetheless, with the caveats just mentioned, it is reasonable to view some uses of SRM as likely to be effective in some plausible future scenarios. Importantly, depending on the pessimistic scenario in question, SRM could be more effective than many other options. Although mitigation is essential for long-term climate stabilization, it is not likely to be effective in managing short-term risks of injustice, because such risks arise from committed warming due to past emissions. Conversely, SRM has the potential to cool the climate rapidly, allowing us to manage some of the shorter-term risks that mitigation cannot address. A similar point holds when comparing the effectiveness of SRM to that of (politically feasible varieties of) adaptation or CDR. In order to be effective in managing short-term risks, any effective adaptation or CDR policy would likely be very expensive and require a rapid scaling-up. Even if they are technically feasible, effective varieties of CDR or adaptation are unlikely to be politically feasible, as parties with the capacity to pay for these expensive policies may be unwilling to do so, at least not at the rapid pace that might be required in a pessimistic scenario.

Finally, let us turn to the issue of moral permissibility. As we saw in chapter two, SRM has the potential to reduce various risks of emissions-related injustice. I will not repeat myself here, aside from noting that this makes it reasonable to ask whether, in pessimistic scenarios, SRM might help some hybrid climate policy satisfy the moral permissibility condition of non-ideal justice. As we saw in chapter two, it is possible to imagine uses of SRM that are plausibly taken to be morally permissible, but such uses seemed to undercut the political feasibility of SRM. So there is likely to be tension between satisfying the political feasibility condition and satisfying the moral permissibility condition when it comes to crafting an SRM policy. As I also noted in chapter two, the political feasibility of SRM may hinge on the fact that it could be deployed by a relatively small group of actors, if for no other reason than that it would be easier for a small group than a large one to agree on objectives for SRM, such as temperature targets, as well as its technical design, such as location of deployment. All else being equal, however, considerations of moral permissibility would favor broad participation in such decision-making about SRM, including all parties who have a legitimate claim to do so (via legitimate representatives). As we saw in chapter three, unilateral or small-scale multilateral decision-making on global climate policy plausibly involves procedural injustice, for it excludes many parties with legitimate claims on decision-making. This might also lead to forms of distributive injustice to present and future parties due to the possibility of parochial decisions of small groups. To address this problem, we might require that (non-ideally just) forms of SRM include broad participation in decision-making. But this move threatens to undermine the political feasibility of SRM, because it may be very difficult for a large group—say, all parties covered by the UNFCCC—to reach an agreement on how to proceed. This is especially the case if we suppose that procedural justice requires consensus on such matters, but this concern remains even if we rely on non-ideal notions of procedural justice, such as one employing the idea of incomplete fairness discussed in chapter two. In short, one might object that SRM is unlikely to satisfy the requirements of non-ideal justice, because meeting one condition thereof makes it likely that another condition thereof will be violated.

This objection loses much of its force in pessimistic climate scenarios, however. In such cases, we must choose a course of action that carries substantial moral problems, for the simple reason that all feasible courses of action carry substantial moral problems, including the option to do nothing. I have argued that we should address pessimistic climate scenarios by identifying the politically feasible policy options that are likely to be effective in achieving their ends and comparing them in light of their moral merits and deficiencies. As a matter of non-ideal justice, we ought then to choose that feasible and effective policy option that is, from a moral point of view, better than any of the other feasible and effective policy options. As I noted above, the moral permissibility condition of non-ideal justice needs to be utilized in a slightly different way here than it is utilized in non-pessimistic scenarios. Normally, the fact that some course of action fails to be proportionate is sufficient to render it unjust in the non-ideal sense. That is, if some policy imposes moral ills while failing to secure proportionate moral goods, then (ordinarily) we may view this policy as non-ideally unjust in the deontic sense. Depending on the nature of a given pessimistic scenario, however, it may be that none of the feasible and effective options are likely to secure properly proportionate ratios of ills to goods. Unless we accept that such scenarios are genuine moral dilemmas—in which case any available option would be morally impermissible—we should not treat proportionality as a necessary condition of non-ideal justice in cases of this kind. Rather, we should focus on the comparative component of the permissibility condition, for the question is essentially a comparative one, namely which available policy is best in an all-things-considered moral sense.

We might worry that this would license the adoption of disproportionate policies in pessimistic climate scenarios. That is true, but it would do so only in those cases in which proportionality cannot be achieved, at least not by policies that are feasible and effective. Nonetheless, we are not to ignore considerations of proportionality even in those unfortunate scenarios. In comparing the relative merits and deficiencies of various options, we will attend to the respective ratios of moral ills to good that each policy is likely to entail. Supposing that all of these ratios involve disproportionate measures of moral ills, some of these ratios are likely to be worse than others. Here we would have good moral reason to choose that disproportionate option that is least bad. Although this option would fail to satisfy the proportionality criterion that usually holds for non-ideal justice, it nonetheless does better (or less bad) than all other available options when it comes to the issue of proportionality. This is to say that such a policy does well on the comparative criterion of non-ideal justice. Coupled with the fact (when it is a fact) that proportionality is not possible, it is plausible to deem the comparatively best policy to be non-ideally just.

With this in mind, we also can see how it is plausible that SRM might be non-ideally just in some pessimistic scenarios, despite the tension between political feasibility and moral permissibility. In ordinary, non-pessimistic scenarios, it may be that this tension prevents SRM policies from qualifying as non-ideally just. If so, then we ought to prefer some politically feasible policy that is likely to be effective, imposes moral ills that are proportionate to the moral goods it delivers, and compares well (in a moral sense) to other politically feasible options that are likely to be effective. But scenarios of the ordinary, non-pessimistic kind are not my focus here. Rather, the question is how we ought to proceed in a pessimistic scenario, specifically one in which no course of action is able to satisfy the proportionality criterion as it applies in ordinary cases. A policy involving SRM might be deeply problematic. In order to be politically feasible, this policy might sacrifice a great deal of procedural fairness. This is a substantial moral ill on the part of that policy, and it might prevent that policy from satisfying the proportionality criterion. If there is some other politically feasible option that is likely to be effective and succeeds in being proportionate, then we ought to prefer that policy to the one involving SRM. But if we are dealing with a case in which proportionality is impossible, we must ask how a disproportionate SRM policy compares to the disproportionate non-SRM policies that satisfy both the feasibility and effectiveness conditions. We have observed some reasons for thinking that a (hybrid) SRM policy might fare better than these other options in certain pessimistic scenarios. Here we would be looking to adopt the least bad option available. Although disproportionate SRM would carry axiological injustices, deploying it would be deontically just (in the non-ideal sense) if each of the other options would carry even worse axiological injustices. It is very obvious that humanity ought to avoid walking into scenarios in which this would be the case. Unfortunately, that may be precisely what we are doing.

Non-Virtuous Decision-Makers and Non-Ideal Justice

There is an important objection to the position I have defended in this chapter, similar to (yet distinct from) the "Realism Objection" discussed in chapter four. The present objection runs as follows. Let it be granted that there is available some non-ideally just (in the deontic sense) use of SRM in a given pessimistic scenario. Despite its availability, actual decision-makers are unlikely to use SRM in this non-ideally just fashion. Rather, if they utilize SRM at all, it is likely to be in ways that run counter to non-ideal justice. This is because those decision-makers are likely to be non-virtuous. For instance, they might lack the virtue of justice and thus fail to respond (reliably) to considerations of non-ideal justice, or they might lack practical wisdom and thus fail to see how SRM ought to be used under the pessimistic circumstances that might hold. More worrisome still, such decision-makers might be vicious, with their decisions being guided by greedy, unjust, short-sighted, or hubristic considerations. To motivate this expectation that likely decision-makers would be non-virtuous (if not vicious), we need only look to human history, replete with examples of injustice and incompetence in matters of war and oppression of various social groups, to take just two examples. If virtuous decision-makers have been rare in history—or if, despite being present, their influence has been slight due to countervailing social factors—then we have no reason to expect decisions in pessimistic climate scenarios to be consonant with virtue. Accordingly, likely decision-makers cannot be counted on to respond (reliably) in appropriate ways to the fact (if and when it is a fact) that non-ideal justice favors some use of SRM. Instead, if they utilize SRM at all, it is likely to be in ways that are (by the lights of non-ideal theory) either ineffectual or impermissible.

What should we say in response to this objection? Let us begin with some preliminary observations. First, it should be noted that the objection calls into question the very point of appealing to non-ideal climate justice. Doing so was supposed to allow us to tap into policies that, by being responsive to social realities (e.g., the fact that some powerful parties will refuse to comply with duties of ideal justice), are practicable. If the objection goes through, then this is not the case, as there would be no reliable connection between non-ideally just policies and those policies that are likely to be adopted. This threatens to undermine the value of non-ideal theory as applied to climate policy. If it turns out that non-ideally just climate policies are—like their ideally just counterparts—unlikely to be utilized in the real world, then what practical value does non-ideal climate justice carry? If we are going to advocate for policies that are unlikely to be adopted, presumably we should advocate for those unlikely policies that are morally preferable, or those which satisfy the requirements of ideal justice.

Second, one of the responses I gave to the Realism Objection will not help here. In chapter four I noted that, if virtuous agents of SRM are unlikely, this gives us a *pro tanto* reason to oppose SRM deployment. This would be useful information, so virtue-theoretic considerations of SRM can be practically useful, even if they only show us that virtuous SRM is very unlikely. This response will not work at present, however, because the objection in question states that because decision-makers are unlikely to be virtuous, it is the case that non-ideally just varieties of SRM are unlikely to be pursued. This objection calls into question the practical value of non-ideal theory. Pointing out that we have a *pro tanto* ethical reason to oppose SRM deployment will not by itself save non-ideal theoretic approaches from this attack. This is because the targets of the Realism Objection and the current objection are distinct.

Third, we should be very clear that the concern here is about decision-makers themselves, not researchers or advocates of SRM. We cannot refute the objection by pointing out that proponents of SRM research are concerned about the moral issues raised by climate change and interested in SRM's potential to ameliorate some of those issues.[22] Proponents of the objection might simply note that actual (non-virtuous) decision-makers are free to ignore the preferences of (virtuous) researchers, including their preferences for non-ideally just climate policies. After all, if decision-makers had been heeding the (sometimes explicitly moral) concerns voiced by climate scientists over the past decades, we likely would have seen the adoption of ambitious mitigation policies some time ago.

Nonetheless, the objection misses its mark. There is an important difference between what is politically *feasible* and what is politically *likely*. To claim that some climate policy is non-ideally just involves the claim that the policy in question is feasible. It is not the business of non-ideal theory to make claims about what policies are politically likely, nor to advocate exclusively on behalf of politically likely options. An essential feature of non-ideal justice is that it has some normative force. That is, it specifies how we ought (morally) to act, and there is no guarantee that how we ought to act will turn out to be either easy or likely. If we limit ourselves only to considering those options that are likely to be taken (e.g., by non-virtuous decision-makers), we risk effectively endorsing a kind of quietism. Whatever that project would be, it is not recognizably a matter of justice to limit ourselves in this way. Perhaps this clarifies an important quality of non-ideal justice. Non-ideal theory limits itself to those options that are feasible given social and political realities. However, within this constraint (plus that of effectiveness), we are obligated to adopt policies that are morally permissible, which could be fairly demanding depending on the situation. In the case of pessimistic climate scenarios, I have argued that we are (non-ideally) obligated to choose that feasible, effective policy that is best in an all-things-considered moral sense. The adoption of such a policy might be politically unlikely, but that does not speak against the value of non-ideal theory, the task of which is merely to pick out which options are non-ideally just under the circumstances that happen to prevail. If we fail to take the (potentially difficult but still feasible) course of action we ought to take, that speaks to our collective moral failure rather than to some inadequacy of non-ideal theory.

Taking Regret Seriously

Suppose that, in the future, we do find ourselves in a pessimistic climate scenario. Suppose further that a policy involving some type of SRM is, morally speaking, our best feasible and effective option in that case. I have argued here that, in such a case, we would be permitted—indeed, obligated—to deploy this type of SRM, despite the substantial moral problems that it is likely to carry. For the reasons I gave in chapter five, it would not be satisfying to view this pessimistic scenario as a genuine moral dilemma, or a case in which all courses of action would be morally impermissible, because this would undermine moral action-guidance precisely when it would be most needed. At the same time, we should not ignore the deep moral problems that deploying SRM might entail, including risks of injustice to innocent parties (e.g., low emitters) and to future generations. It is appropriate to acknowledge these deep moral problems in a serious way, genuinely recognizing both the axiological injustices that SRM might bring about and the enormous moral failure (on the part of some) indicated by the very existence of a pessimistic climate scenario that calls for SRM. Regarding the latter, it is very unlikely that such a pessimistic scenario would have arisen had many of us not shirked our past obligations to mitigate our emissions. As I argued in chapter five, even if SRM is morally permissible (and, further, non-ideally just in the deontic sense), the agents of SRM should feel agent-regret for deploying it.

It might be easy to overlook the importance of agent-regret here, but taking this attitude seriously is actually very important. To be clear, I am not suggesting that harboring agent-regret is a necessary condition for deployment of SRM to be non-ideally just. These two issues are distinct. Whether some policy satisfies the demands of non-ideal justice depends upon that policy's feasibility, likely effectiveness, and moral permissibility. In a given case, some policy involving SRM might satisfy these demands even if proponents of that policy feel no agent-regret in deploying SRM. Nonetheless, although it would not change the fact (if it is one) that SRM is non-ideally just in this case, it would be a bad thing for those deploying SRM to lack agent-regret in doing so. There are two reasons for thinking this would be bad. First, it seems morally inappropriate (but not necessarily wrong) not to regret the potential moral ills of a policy one is enacting. The same holds for not regretting one's part in some historical moral failure that creates a situation in which our best option is morally problematic. Someone who fails to regret such matters is rather like the Satisfied Climate Engineer we encountered in chapter five, ignoring the moral disvalue of SRM, including the moral disvalue of its likely effects on others and its impact on one's own life. Intuitively, it matters how we regard our own actions and the circumstances that surround them, even in cases in which those actions are blameless. But those who enact an SRM policy while eschewing agent-regret run a dangerous risk of ignoring such morally salient features. By failing to regret their role in bringing about a morally disvaluable outcome, for example, they do not fully acknowledge something that appears to matter a great deal, namely that *their own* actions are likely to bring about morally bad states of affairs. To adapt Williams' comment regarding the blameless truck driver who kills a child but shrugs off any regret regarding *his own* role in that outcome, most of us are likely to feel some doubt about such a person, despite the fact (if it is one) that her actions meet the requirements of non-ideal justice. This gives us a reason to think that agent-regret for SRM is morally appropriate in its own right.

The second reason for favoring agent-regret with respect to non-ideally just SRM policies is a pragmatic one: actors who take agent-regret seriously are, all else being equal, more likely than actors who eschew agent-regret to deploy and maintain SRM in morally defensible ways. Parties who take on agent-regret for their actions (and the conditions thereof) are likely to be attuned to the moral salience of various, relevant features of the case. For instance, if I am to regret my helping to bring about risks of (axiological) injustice to low emitters or future generations, I must attend both to what those risks involve and to the fact of the moral disvalue they portend. If I am not attuned to such things, then it is likely that I will do a poor job of regretting my role in bringing about risks of injustice. In that case, although I might have a vague sense of wishing things had been otherwise, I would not be taking agent-regret seriously. For one thing, I would not really be apprehending what is bad about the risks imposed by SRM, precisely because (in our example) I would not be paying those risks much attention. This will prevent me from appreciating what is morally salient about them—for instance, what those risks would mean in concrete terms for actual persons, were they to be realized. So if I am serious about agent-regret, I will attend to such details. In that sense, I will be attuned to the morally salient features of an SRM policy, including the context of its deployment, as well as its likely impacts. All else being equal, an agent of SRM who does not take on agent-regret will likely be less attuned to such features. At the very least, such attunement will be much more reliable in those who are serious about agent-regret. Such persons are thus likely to be more adept at seeing what features matter (morally) when it comes to deploying and maintaining SRM.

Moreover, it is plausible to think that those feeling agent-regret will be more motivated than their counterparts to maintain (and improve) an SRM policy in morally relevant ways, such as by reducing risks of injustice that become apparent only after deployment. Because regret involves wishing something had been otherwise, it is inherently motivational. If I genuinely wish that axiological injustices were avoidable, for instance, then to some extent I will be moved to avoid or reduce those injustices if doing so becomes possible. Because of the far-reaching (and often deep) uncertainty in our understanding of natural and social systems that are relevant to climate policy-making, we may discover that it is indeed possible to improve our chosen climate policy even after it has been enacted. Accordingly, we should want the overseers of that policy to be both motivated to improve it when possible and to be skilled in identifying ways of doing so. Because, relative to its absence, agent-regret is more likely to indicate such motivation and skill in the persons who harbor it, this gives us another reason to want agent-regret to accompany deployment of non-ideally just SRM. This mirrors a claim I defended in chapter four, namely that virtuous agents are, all else being equal, more likely than non-virtuous ones to deploy and maintain SRM in ways that are both morally permissible and likely to yield morally valuable (or minimize morally disvaluable) outcomes. Indeed, we might think that the fully virtuous person necessarily would feel agent-regret in enacting a non-ideally just climate policy. A person with the virtues of justice and benevolence, for example, genuinely cares about those goods and wishes to see them prevail. If it is in fact infeasible for those goods to prevail in full, such a person will judge this to be a bad thing. Plausibly, she will wish that things had been otherwise, regretting that the best (or least bad) moral option involves taking part in a problematic policy.

Now one might object to this by pointing out that we simply do not have time to ensure that the agents behind an SRM policy are friendly to agent-regret. Much as we cannot afford to demand ideal justice in our non-ideal world, so we cannot afford to demand ideal agents for non-ideal policies. Rather, as one might argue, we should take the best agents that we can get, their likely imperfections notwithstanding. But this objection would misunderstand the role agent-regret is playing in my account. My claim is only that it is good to harbor (and bad to lack) agent-regret for deploying SRM. Once again, I am not treating agent-regret as a necessary condition for the moral permissibility of SRM, nor for its being (non-ideally) just more generally. Indeed, it might be the case that, in some pessimistic scenario, the morally best policy involves SRM as deploying by non-regretting parties. This lack of regret would not change the fact that, on my account, SRM ought to be deployed in this case. Nonetheless, we should acknowledge that this lack of regret is morally inappropriate, for the reasons given above. Further, it would be reasonable for us to be suspicious that these non-regretting agents might not fully understand the morally salient features that are in play, again for the reasons given above. This is not to call into question the motives of those parties, but rather to express doubt that they will succeed in their (let us assume) sincere goal of pursuing the best option in a pessimistic scenario. Despite good intentions, their lack of agent-regret might signal a lack of moral attunement to what matters, and this might lead them to make mistakes when it comes to deploying and maintaining SRM in the morally best fashion. So we have two distinct reasons to want agents of SRM to feel agent-regret, but this does not entail that agents of SRM must be open to agent-regret before it may be deployed, and so my stance is not vulnerable to the present objection.

Throughout this book, I have been relying on a distinction between what is good and what ought to be done. It may be that some variety of SRM ought to be deployed in the future, but it would be bad if things came to that.

Notes

1. Parts of this chapter are adapted from Toby Svoboda, "Aerosol Geoengineering Deployment and Fairness," *Environmental Values* 25, no. 1 (2016): 51–68.
2. David Keith, *A Case for Climate Engineering* (Cambridge: The MIT Press, 2013).
3. Toby Svoboda, "Is Aerosol Geoengineering Ethically Preferable to Other Climate Change Strategies?" *Ethics & the Environment* 17, no. 2 (2012): 111–35.
4. W. Neil Adger et al., "Are There Social Limits to Adaptation to Climate Change?," *Climatic Change* 93, nos. 3–4 (2009): 335–54.
5. Juan B. Moreno-Cruz and David W. Keith, "Climate Policy under Uncertainty: A Case for Solar Geoengineering," *Climatic Change* 121, no. 3 (2013): 431–44.
6. Christopher J. Preston, "Climate Engineering and the Cessation Requirement: The Ethics of a Life-Cycle," *Environmental Values* 25, no. 1 (2016): 91–107.
7. Jason J. Blackstock et al., "Climate Engineering Responses to Climate Emergencies," http://arxiv.org/pdf/0907.5140.
8. Joshua Horton, "The Emergency Framing of Solar Geoengineering: Time for a Different Approach," *The Anthropocene Review* 2, no. 2 (2015), 147–151.
9. IPCC, *Climate Change 2014: Mitigation of Climate Change: Working Group III Contribution to the IPCC Fifth Assessment Report* (Cambridge: Cambridge University Press, 2015); National Research Council, *Climate Intervention: Reflecting Sunlight to Cool Earth* (Washington, D.C.: National Academy of

Sciences, 2015); John G. Shepherd et al., *Geoengineering the Climate: Science, Governance and Uncertainty* (London: Royal Society, 2009).

10. Arctic Methane Emergency Group, "AMEG's Declaration," http://ameg.me/index.php.
11. Jim M. Haywood et al., "Asymmetric Forcing from Stratospheric Aerosols Impacts Sahelian Rainfall," *Nature Climate Change* 3, no. 7 (2013): 660.
12. Peter J. Irvine, A. Ridgwell, and D. J. Lunt, "Assessing the Regional Disparities in Geoengineering Impacts," *Geophysical Research Letters* 37 (2010).
13. Keith, *A Case for Climate Engineering.*
14. T. M. L. Wigley, "A Combined Mitigation/Geoengineering Approach to Climate Stabilization," *Science* 314, no. 5798 (2006): 452–4.
15. Joshua Horton, "Geoengineering and the Myth of Unilateralism: Pressures and Prospects for International Cooperation," *Stanford Journal of Law, Science & Policy* 4 (2011): 56–69.
16. Simon Caney, "Climate Change and Non-Ideal Theory: Six Ways of Responding to Non-Compliance," in *Climate Justice in a Non-Ideal World,* ed. Clare Heyward and Dominic Roser (Oxford: Oxford University Press, 2016), 21–42.
17. Nicholas Apergis and James E. Payne, "Renewable Energy Consumption and Economic Growth: Evidence from a Panel of OECD Countries," *Energy Policy* 38, no. 1 (2010): 656–60.
18. Benjamin Hale, "The World That Would Have Been: Moral Hazard Arguments against Geoengineering," in *Engineering the Climate: The Ethics of Solar Radiation Management,* ed. Christopher J. Preston (Lanham: Lexington Books, 2012), 113–31; Jesse Reynolds, "A Critical Examination of the Climate Engineering Moral Hazard and Risk Compensation Concern," *The Anthropocene Review* 2, no. 2 (2014), 174–191.
19. David J. Houston and Lilliard E. Richardson, "Risk Compensation or Risk Reduction? Seatbelts, State Laws, and Traffic Fatalities," *Social Science Quarterly* 88, no. 4 (2007): 913–36.
20. David Keith, Edward Parson, and M. Granger Morgan, "Research on Global Sun Block Needed Now," *Nature* 463, no. 7280 (2010): 426–27.
21. Ibid.
22. Joshua Horton and David Keith, "Solar Geoengineering and Obligations to the Global Poor," in *Climate Justice and Geoengineering: Ethics and Policy in the Atmospheric Anthropocene,* ed. Christopher Preston (Lanham: Rowman & Littlefield International, 2016), 79–92.

Bibliography

Adger, W. Neil, Suraje Dessai, Marisa Goulden, Mike Hulme, Irene Lorenzoni, Donald R. Nelson, Lars Otto Naess, Johanna Wolf, and Anita Wreford. "Are There Social Limits to Adaptation to Climate Change?" *Climatic Change* 93, nos. 3–4 (2009): 335–54.

Allen, Myles. "Liability for Climate Change." *Nature* 421, no. 6926 (2003): 891–92.

Apergis, Nicholas, and James E. Payne. "Renewable Energy Consumption and Economic Growth: Evidence from a Panel of OECD Countries." *Energy Policy* 38, no. 1 (2010): 656–60.

Arctic Methane Emergency Group. "AMEG's Declaration." http://ameg.me/index.php.

Arneson, Richard J. "Equality and Equal Opportunity for Welfare." *Philosophical Studies* 56, no. 1 (1989): 77–93.

Asilomar Scientific Organizing Committee. *The Asilomar Conference Recommendations on Principles for Research into Climate Engineering Techniques.* Washington, D.C.: Climate Institute, 2010.

Baatz, Christian. "Responsibility for the Past? Some Thoughts on Compensating Those Vulnerable to Climate Change in Developing Countries." *Ethics, Policy & Environment* 16, no. 1 (2013): 94–110.

Baatz, Christian, and Konrad Ott. "Why Aggressive Mitigation Must Be Part of Any Pathway to Climate Justice." In *Climate Justice and Geoengineering: Ethics and Policy in the Atmospheric Anthropocene*, edited by Christopher Preston, 93–108. Rowman & Littlefield International, 2016.

Barrett, Scott. "The Incredible Economics of Geoengineering." *Environmental and Resource Economics* 39, no. 1 (2008): 45–54.

Barrett, Scott, Timothy M. Lenton, Antony Millner, Alessandro Tavoni, Stephen Carpenter, John M. Anderies, F. Stuart Chapin III, et al. "Climate Engineering Reconsidered." *Nature Climate Change* 4, no. 7 (2014): 527–29.

Bickel, J. Eric, and Lee Lane. "An Analysis of Climate Engineering as a Response to Climate Change." Frederiksberg: Copenhagen Consensus Center, 2009.

Blackstock, Jason J., David Battisti, Ken Caldeira, Douglas E. Eardley, Jonathan I. Katz, David W. Keith, Steven E. Koonin, Aristides A. N. Patrinos, Daniel P. Schrag, and Robert H. Socolow. "Climate Engineering Responses to Climate Emergencies." http://arxiv.org/pdf/0907.5140.

Bradley, Richard, and Katie Steele. "Making Climate Decisions." *Philosophy Compass* 10, no. 11 (2015): 799–810.

Broome, John. *Climate Matters: Ethics in a Warming World.* New York: W.W. Norton & Company, 2012.

———. "Discounting the Future." *Philosophy & Public Affairs* 23, no. 2 (1994): 128–56.

———. "Fairness." *Proceedings of the Aristotelian Society* 91 (1990): 87–101.

Bunzl, Martin. "Geoengineering Harms and Compensation." *Stanford Journal of Law, Science & Policy* 4 (2011): 70–76.

Caney, Simon. "Climate Change and Non-Ideal Theory: Six Ways of Responding to Non-Compliance." In *Climate Justice in a Non-Ideal World*, edited by Clare Heyward and Dominic Roser, 21–42. Oxford: Oxford University Press, 2016.

———. "Cosmopolitan Justice, Responsibility, and Global Climate Change." *Leiden Journal of International Law* 18, no. 4 (2005): 747–75.

Carson, Rachel. *Silent Spring*. Boston: Houghton Mifflin, 2002.

Clingerman, Forrest. "Between Babel and Pelagius: Religion, Theology, and Geoengineering." In *Engineering the Climate: The Ethics of Solar Radiation Management*, edited by Christopher Preston, 201–20. Lanham: Lexington Books, 2012.

Clingerman, Forrest, and Kevin J. O'Brien. "Playing God: Why Religion Belongs in the Climate Engineering Debate." *Bulletin of the Atomic Scientists* 70, no. 3 (2014): 27–37.

Colyvan, Mark, Damian Cox, and Katie Steele. "Modelling the Moral Dimension of Decisions." *Noûs* 44, no. 3 (2010): 503–29.

Corner, Adam, Karen Parkhill, Nick Pidgeon, and Naomi E. Vaughan. "Messing with Nature? Exploring Public Perceptions of Geoengineering in the UK." *Global Environmental Change* 23, no. 5 (2013): 938–47.

Crutzen, Paul J. "Albedo Enhancement by Stratospheric Sulfur Injections: A Contribution to Resolve a Policy Dilemma?" *Climatic Change* 77, nos. 3–4 (2006): 211–19.

Dennig, Francis, Mark B. Budolfson, Marc Fleurbaey, Asher Siebert, and Robert H. Socolow. "Inequality, Climate Impacts on the Future Poor, and Carbon Prices." *Proceedings of the National Academy of Sciences* 112, no. 52 (2015): 15827–32.

Dworkin, Ronald. "What Is Equality? Part 1: Equality of Welfare." *Philosophy & Public Affairs* 10, no. 3 (1981): 185–246.

———. "What Is Equality? Part 2: Equality of Resources." *Philosophy & Public Affairs* 10, no. 4 (1981): 283–345.

Estlund, David. "Utopophobia." *Philosophy & Public Affairs* 42, no. 2 (2014): 113–34.

Ferraro, Angus J., Eleanor J. Highwood, and Andrew J. Charlton-Perez. "Weakened Tropical Circulation and Reduced Precipitation in Response to Geoengineering." *Environmental Research Letters* 9, no. 1 (2014).

Fleming, James R. *Fixing the Sky: The Checkered History of Weather and Climate Control*. New York: Columbia University Press, 2010.

Garcia, Robert K. "Towards a Just Solar Radiation Management Compensation System: A Defense of the Polluter Pays Principle." *Ethics, Policy & Environment* 17, no. 2 (2014): 178–82.

Gardiner, Stephen M. "Ethics and Global Climate Change." *Ethics* 114, no. 3 (2004): 555–600.

———. "Ethics, Geoengineering and Moral Schizophrenia: What's the Question?" In *Climate Change Geoengineering: Philosophical Perspectives, Legal Issues, and Governance Frameworks*, edited by William C.G. Burns and Andrew Strauss, 11–38. Cambridge: Cambridge University Press, 2013.

———. "Is 'Arming the Future' with Geoengineering Really the Lesser Evil? Some Doubts about the Ethics of Intentionally Manipulating the Climate System." In

Climate Ethics, edited by Stephen M. Gardiner, Simon Caney, Dale Jamieson, and Henry Shue, 284–312. New York: Oxford University Press, 2010.

———. "A Perfect Moral Storm: Climate Change, Intergenerational Ethics and the Problem of Moral Corruption." *Environmental Values* 15, no. 3 (2006): 397–413.

———. *A Perfect Moral Storm: The Ethical Tragedy of Climate Change*. Oxford: Oxford University Press, 2011.

Goes, Marlos, Nancy Tuana, and Klaus Keller. "The Economics (or Lack Thereof) of Aerosol Geoengineering." *Climatic Change* 109, nos. 3–4 (2011): 719–44.

Hale, Benjamin. "The World That Would Have Been: Moral Hazard Arguments against Geoengineering." In *Engineering the Climate: The Ethics of Solar Radiation Management*, edited by Christopher J. Preston, 113–31. Lanham: Lexington Books, 2012.

Hartzell-Nichols, Lauren. "Precaution and Solar Radiation Management." *Ethics, Policy & Environment* 15, no. 2 (2012): 158–71.

Haywood, Jim M., Andy Jones, Nicolas Bellouin, and David Stephenson. "Asymmetric Forcing from Stratospheric Aerosols Impacts Sahelian Rainfall." *Nature Climate Change* 3, no. 7 (2013): 660–5.

Heyward, Clare. "Benefiting from Climate Geoengineering and Corresponding Remedial Duties: The Case of Unforeseeable Harms." *Journal of Applied Philosophy* 31, no. 4 (2014): 405–19.

Heyward, C., and D. Roser. *Climate Justice in a Non-Ideal World*. Oxford: Oxford University Press, 2016.

Honegger, Matthias, Kushini Sugathapala, and Axel Michaelowa. "Tackling Climate Change: Where Can the Generic Framework Be Located." *CCLR* 2 (2013): 125.

Hooker, Brad. "Fairness." *Ethical Theory and Moral Practice* 8, no. 4 (2005): 329–52.

Horton, Joshua. "The Emergency Framing of Solar Geoengineering: Time for a Different Approach." *The Anthropocene Review* 2, no. 2 (2015): 147-51.

———. "Geoengineering and the Myth of Unilateralism: Pressures and Prospects for International Cooperation." *Stanford Journal of Law, Science & Policy* 4 (2011): 56–69.

———. "Solar Geoengineering: Reassessing Costs, Benefits, and Compensation." *Ethics, Policy & Environment* 17, no. 2 (2014): 175-7.

Horton, Joshua, and David Keith. "Solar Geoengineering and Obligations to the Global Poor." In *Climate Justice and Geoengineering: Ethics and Policy in the Atmospheric Anthropocene*, edited by Christopher Preston, 79–92. Lanham: Rowman & Littlefield International, 2016.

Horton, Joshua, Andrew Parker, and David Keith. "Liability for Solar Geoengineering: Historical Precedents, Contemporary Innovations, and Governance Possibilities." *NYU Environmental Law Journal* 22 (2014): 225.

Houston, David J., and Lilliard E. Richardson. "Risk Compensation or Risk Reduction? Seatbelts, State Laws, and Traffic Fatalities." *Social Science Quarterly* 88, no. 4 (2007): 913–36.

Hursthouse, Rosalind. "Normative Virtue Ethics." In *How Should One Live?*, edited by Roger Crisp, 19–33. New York: Oxford University Press, 1996.

———. *On Virtue Ethics*. Oxford: Oxford University Press, 1999.

IPCC [Intergovernmental Panel on Climate Change]. *Climate Change 2014: Mitigation of Climate Change: Working Group III Contribution to the IPCC Fifth Assessment Report*. Cambridge: Cambridge University Press, 2015.

Irvine, Peter J., Andy Ridgwell, and Daniel J. Lunt. "Assessing the Regional Disparities in Geoengineering Impacts." *Geophysical Research Letters* 37, no. 18 (2010).

Irvine, Peter J., Ryan L. Sriver, and Klaus Keller. "Tension between Reducing Sea-Level Rise and Global Warming through Solar-Radiation Management." *Nature Climate Change* 2, no. 2 (2012): 97–100.

Jones, Andy, Jim M. Haywood, Kari Alterskjær, Olivier Boucher, Jason N.S. Cole, Charles L. Curry, Peter J. Irvine, Duoying Ji, Ben Kravitz, and Jón Egill Kristjáns-son. "The Impact of Abrupt Suspension of Solar Radiation Management (Termination Effect) in Experiment G2 of the Geoengineering Model Intercomparison Project (GeoMIP)." *Journal of Geophysical Research: Atmospheres* 118, no. 17 (2013): 9743–52.

Kant, Immanuel. "Groundwork for the Metaphysics of Morals." In *Practical Philosophy*, edited and translated by Mary J. Gregor. New York: Cambridge University Press, 1996.

———. "The Metaphysics of Morals." In *Practical Philosophy*, edited and translated by Mary J. Gregor. New York: Cambridge University Press, 1996.

Keith, David. *A Case for Climate Engineering*. Cambridge: MIT Press, 2013.

———. "Geoengineering the Climate: History and Prospect." *Annual Review of Energy and the Environment* 25 (2000): 245–84.

Keith, David, Edward Parson, and M. Granger Morgan. "Research on Global Sun Block Needed Now." *Nature* 463, no. 7280 (2010): 426–27.

Kraut, Richard. "Aristotle's Ethics." In *The Stanford Encyclopedia of Philosophy*, edited by Edward N. Zalta. Stanford: Metaphysics Research Lab, 2014.

Lamont, Julian, and Christi Favor. "Distributive Justice." In *The Stanford Encyclopedia of Philosophy*, edited by Edward N. Zalta. Stanford: Metaphysics Research Lab, 2016.

Larson, Eric Thomas. "Why Environmental Liability Regimes in the United States, the European Community, and Japan Have Grown Synonymous with the Polluter Pays Principle." *Vanderbilt Journal of Transnational Law* 38 (2005): 541.

Levine, Gabriel Leopold. "'Has It Really Come to This?' An Assessment of Virtue Ethical Approaches to Climate Engineering." Thesis, Yale University, 2014.

McConnell, Terrance C. "Moral Dilemmas and Consistency in Ethics." *Canadian Journal of Philosophy* 8, no. 2 (1978): 269–87.

MacCracken, Michael C. "On the Possible Use of Geoengineering to Moderate Specific Climate Change Impacts." *Environmental Research Letters* 4, no. 4 (2009).

Meyer, Kirsten, and Christian Uhle. "Geoengineering and the Accusation of Hubris." *THESys Discussion Papers*, 2015. www.iri-thesys.org/discussion-papers/paper-pdfs/discussion-paper-2015-3-final.pdf

Mill, John Stuart. *Utilitarianism*, edited by George Sher. Indianapolis: Hackett, 2002.

Millard-Ball, Adam. "The Tuvalu Syndrome." *Climatic Change* 110, nos. 3–4 (2012): 1047–66.

Moellendorf, Darrel. *The Moral Challenge of Dangerous Climate Change: Values, Poverty, and Policy*. Cambridge: Cambridge University Press, 2014.

———. "Taking UNFCCC Norms Seriously." In *Climate Justice in a Non-Ideal World*, edited by Clare Heyward and Dominic Roser, 104–21. Oxford: Oxford University Press, 2016.

Moreno-Cruz, Juan B., and David W. Keith. "Climate Policy under Uncertainty: A Case for Solar Geoengineering." *Climatic Change* 121, no. 3 (2013): 431–44.

Morrow, David R. "Starting a Flood to Stop a Fire: Some Moral Constraints on Solar Radiation Management." *Ethics, Policy & Environment* 17, no. 2 (2014): 123–138.

———. "Ethical Aspects of the Mitigation Obstruction Argument against Climate Engineering Research." *Philosophical Transactions of the Royal Society of London A* 372, no. 2031 (2014).

———. "Fairness in Allocating the Global Emissions Budget." *Environmental Values* (forthcoming).

———. "Wants and Needs in Mitigation Policy." *Climatic Change* 130, no. 3 (2015): 335–45.

Morrow, David R., and Toby Svoboda. "Geoengineering and Non-Ideal Theory." *Public Affairs Quarterly* 30, no. 1 (2016): 85–104.

Morrow, David R., Robert E. Kopp, and Michael Oppenheimer. "Toward Ethical Norms and Institutions for Climate Engineering Research." *Environmental Research Letters* 4, no. 4 (2009).

National Research Council. *Climate Intervention: Carbon Dioxide Removal and Reliable Sequestration*. Washington, D.C.: National Academy of Sciences, 2015.

———. *Climate Intervention: Reflecting Sunlight to Cool Earth*. Washington, D.C: National Academy of Sciences, 2015.

Nolt, John. "Casualties as a Moral Measure of Climate Change." *Climatic Change* 130, no. 3 (2014): 347–358.

Nordhaus, William D. *A Question of Balance: Weighing the Options on Global Warming Policies*. New Haven: Yale University Press, 2008.

Norton, Bryan G. *Toward Unity among Environmentalists*. Oxford: Oxford University Press, 1994.

Nozick, Robert. *Anarchy, State, and Utopia*. New York: Basic Books, 1974.

Nussbaum, Martha C. "Virtue Ethics: A Misleading Category?" *The Journal of Ethics* 3, no. 3 (1999): 163–201.

Ott, Konrad. "Might Solar Radiation Management Constitute a Dilemma?" In *Engineering the Climate: The Ethics of Solar Radiation Management*, edited by Christopher Preston, 33–42. Lanham: Lexington Books, 2012.

Page, Edward. "Intergenerational Justice and Climate Change." *Political Studies* 47, no. 1 (1999): 53–66.

Posner, Eric. "You Can Have Either Climate Justice or a Climate Treaty, Not Both." *Slate*, November 19, 2013. www.slate.com/articles/news_and_politics/view_from_chicago/2013/11/climate_justice_or_a_climate_treaty_you_can_t_have_both.html.

Powys Whyte, Kyle. "Now This! Indigenous Sovereignty, Political Obliviousness and Governance Models for SRM Research." *Ethics, Policy & Environment* 15, no. 2 (2012): 172–87.

Preston, Christopher J. "Climate Engineering and the Cessation Requirement: The Ethics of a Life-Cycle." *Environmental Values* 25, no. 1 (2016): 91–107.

———. "Re-Thinking the Unthinkable: Environmental Ethics and the Presumptive Argument against Geoengineering." *Environmental Values* 20, no. 4 (2011): 457–79.

Rawls, John. *A Theory of Justice*, Revised Edition. Cambridge: Harvard University Press, 1999.

Reynolds, Jesse. "A Critical Examination of the Climate Engineering Moral Hazard and Risk Compensation Concern." *The Anthropocene Review* 2, no. 2 (2014): 174–91.

———. "Why the UNFCCC and CBD Should Refrain from Regulating Solar Climate Engineering." In *Geoengineering Our Climate? Ethics, Politics and Governance*, edited by Jason Blackstock. New York: Routledge, forthcoming.

Ricke, Katharine L., Juan B. Moreno-Cruz, and Ken Caldeira. "Strategic Incentives for Climate Geoengineering Coalitions to Exclude Broad Participation." *Environmental Research Letters* 8, no. 1 (2013).

Sandel, Michael J. *What Money Can't Buy: The Moral Limits of Markets*. New York: Farrar, Straus and Giroux, 2012.

Sen, Amartya. *Choice, Welfare and Measurement*. Cambridge: Cambridge University Press, 1982.

Shepherd, John G., Ken Caldeira, Peter Cox, Joanna Haigh, David W. Keith, Brian Launder, Georgina Mace, Gordon MacKerron, John Pyle, Steve Rayner, Catherine Redgwell, and Andrew Watson. *Geoengineering the Climate: Science, Governance and Uncertainty*. London: Royal Society, 2009.

Shue, Henry. "Global Environment and International Inequality." *International Affairs* 75, no. 3 (1999): 531–45.

Singer, Peter. *The Most Good You Can Do: How Effective Altruism Is Changing Ideas about Living Ethically*. Castle Lectures Series. New Haven: Yale University Press, 2015.

Smith, Richard L., Claudia Tebaldi, Doug Nychka, and Linda O. Mearns. "Bayesian Modeling of Uncertainty in Ensembles of Climate Models." *Journal of the American Statistical Association* 104, no. 485 (2009): 97–116.

Stern, Nicholas. *The Economics of Climate Change: The Stern Review*. Cambridge: Cambridge University Press, 2007.

Stone, Daithi A., Myles R. Allen, Peter A. Stott, Pardeep Pall, Seung-Ki Min, Toru Nozawa, and Seiji Yukimoto. "The Detection and Attribution of Human Influence on Climate." *Annual Review of Environment and Resources* 34 (2009): 1–16.

Stott, Peter A., D. A. Stone, and M. R. Allen. "Human Contribution to the European Heatwave of 2003." *Nature* 432, no. 7017 (2004): 610–14.

Sunstein, Cass R. *Laws of Fear: Beyond the Precautionary Principle*. Cambridge: Cambridge University Press, 2005.

Svoboda, Toby. "Aerosol Geoengineering Deployment and Fairness." *Environmental Values* 25, no. 1 (2016): 51–68.

———. "Geoengineering, Agent-Regret, and the Lesser of Two Evils Argument." *Environmental Ethics* 37, no. 2 (2015): 207–20.

———. "Is Aerosol Geoengineering Ethically Preferable to Other Climate Change Strategies?" *Ethics & the Environment* 17, no. 2 (2012): 111–35.

———. "Solar Radiation Management and Comparative Climate Justice." In *Climate Justice and Geoengineering: Ethics and Policy in the Atmospheric Anthropocene*, edited by Christopher Preston, 3–14. Lanham: Rowman & Littlefield International, 2016.

Svoboda, Toby, and Peter J. Irvine. "Ethical and Technical Challenges in Compensating for Harm Due to Solar Radiation Management Geoengineering." *Ethics, Policy & Environment* 17, no. 2 (2014): 157–74.

Svoboda, Toby, Klaus Keller, Marlos Goes, and Nancy Tuana. "Sulfate Aerosol Geoengineering: The Question of Justice." *Public Affairs Quarterly* 25, no. 3 (2011): 157–80.

Swartwood, Jason D. "Wisdom as an Expert Skill." *Ethical Theory and Moral Practice* 16, no. 3 (2013): 511–28.

Tilmes, Simone Rolf Muller, and Ross Salawitch. "The Sensitivity of Polar Ozone Depletion to Proposed Geoengineering Schemes." *Science* 320, no. 5880 (2008): 1201–4.

Tuana, Nancy, Ryan L. Sriver, Toby Svoboda, Roman Olson, Peter J. Irvine, Jacob Haqq-Misra, and Klaus Keller. "Towards Integrated Ethical and Scientific Analysis of Geoengineering: A Research Agenda." *Ethics, Policy & Environment* 15, no. 2 (2012): 136–57.

United Nations Framework Convention on Climate Change. New York: 1992. http://unfccc.int/files/essential_background/background_publications_htmlpdf/application/pdf/conveng.pdf

Victor, David G. "On the Regulation of Geoengineering." *Oxford Review of Economic Policy* 24, no. 2 (2008): 322–36.

Victor, David G., M. Granger Morgan, Jay Apt, John Steinbruner, and Katharine Ricke. "The Geoengineering Option: A Last Resort against Global Warming?" *Foreign Affairs* 88, no. 2 (2009): 64–76.

Weitzman, Martin L. "A Review of the Stern Review on the Economics of Climate Change." *Journal of Economic Literature* 45, no. 3 (2007): 703–24.

Wiens, David. "Prescribing Institutions without Ideal Theory." *Journal of Political Philosophy* 20, no. 1 (2012): 45–70.

Wigley, T. M. L. "A Combined Mitigation/Geoengineering Approach to Climate Stabilization." *Science* 314, no. 5798 (2006): 452–54.

Williams, Bernard. *Ethics and the Limits of Philosophy.* Cambridge: Harvard University Press, 1985.

Wingspread Statement. "Precautionary Principle," 1998. www.sehn.org/wing.html.

Wong, Pak-Hang. "Confucian Environmental Ethics, Climate Engineering, and the 'Playing God' Argument." *Zygon* 50, no. 1 (2015): 28–41.

Zimmerman, Michael J. *The Concept of Moral Obligation.* New York: Cambridge University Press, 1996.

Index

Printed and bound by CPI Group (UK) Ltd, Croydon, CR0 4YY

24/10/2024

01778282-0005